D0758816

ENERGY FROM HEAVEN AND EARTH

DISTINGUISHED LECTURE SERIES, STATE UNIVERSITY OF NEW YORK AT BUFFALO

In 1969, the Faculty of Natural Sciences and Mathematics at the State University of New York at Buffalo established the Distinguished Visiting Lectureship of the Faculty of Natural Sciences and Mathematics. This lectureship is sponsored by the State University of New York at Buffalo and awarded annually to an outstanding scientist, who presents a series of four lectures at the University. These lectures are to cover as wide a range of material as possible, not only recent developments in research, but also the more philosophical aspects of particular areas of endeavor. The lectures are also intended to stimulate and encourage an interest in the sciences in as wide an audience as possible, among both the University and the Buffalo communities. This monograph is partly based on the 1975 lectures of Dr. Edward Teller.

Distinguished Visiting Lecturers of the Faculty of Natural Sciences and Mathematics at the State University of New York at Buffalo

1970 Linus Pauling

1971 Fred Hoyle

1972 George Wald

1973 Bernard Lovell

1974 Henry Eyring

1975 Edward Teller

HARVEY PRIZE, TECHNION (ISRAEL INSTITUTE OF TECHNOLOGY AT HAIFA)

In 1971, the Harvey Prize was established by Leo M. Harvey, the founder of the Harvey Aluminum Company. The prize was conceived with three primary goals: to reward excellence; to make known throughout the world the singular achievement of the prize winners in their respective fields; and to enable the Technion and other institutions of higher learning in Israel to enjoy their presence and inspiration. The prize is awarded annually in one or more of the following fields: science and technology; human health; advancement of peace in the Middle East; literature of profound insight into the life of the peoples of the Middle East. Recipients of the prize travel and lecture in Israel for a period of two weeks. This monograph is partly based on the 1975 lectures of Dr. Edward Teller.

Harvey Prize Recipients in Science and Technology

1972 Claude E. Shannon—information theory

1973 Prizes postponed to 1974 because of Yom Kippur War

1974 Allan Howard Cottrell—mechanical properties of materials

1975 Edward Teller—nuclear science and solid state physics

1976 Herman F. Mark—polymers and plastics

1977 Seymour Benzer—molecular and behavioral genetics

1977 Freeman J. Dyson—quantum electrodynamics and ferromagnetism

1978 Isaak Wahl—techniques to improve cereal grains

ENERGY
FROM HEAVEN AND EARTH

EDWARD TELLER

In which a story is told about energy from its origins 15,000,000,000 years ago to its present adolescence —turbulent, hopeful, beset by problems and in need of help.

W. H. FREEMAN AND COMPANY
San Francisco

Sponsoring Editor: Peter Renz
Project Editor: Pearl C. Vapnek
Manuscript Editor: Suzanne Lipsett
Designer: Marie Carluccio
Production Coordinator: Chuck Pendergast
Illustration Coordinator: Cheryl Nufer
Composition: Typesetting Services of California
Printer and Binder: The Maple-Vail Book Manufacturing Group

Library of Congress Cataloging in Publication Data

Teller, Edward, 1908–
Energy from heaven and earth.

Based on lectures given by the author during 1975 for the Harvey Prize Lectures at Technion, Haifa, Israel; the Distinguished Visiting Lectures of the Faculty of Natural Sciences and Mathematics, State University of New York at Buffalo; and a lecture delivered at Acadia University, Wolfville, Nova Scotia.

1. Power resources. I. Title.
TJ163.2.T4 333.7 79-4049
ISBN 0-7167-1063-3
ISBN 0-7167-1064-1 pbk.

Printed in the United States of America

9 8 7 6 5 4 3 2

This book is dedicated to
the memory of Nelson A. Rockefeller,
whose advice and inspiration
aided me in writing this book,
and whose vision and initiative,
had they only been followed,
would have made this book
unnecessary.

CONTENTS

and the need to invigorate research, we face the impossible task of deciding what we do next. We may assume that if we disregard the energy problem it will go away; or we may decide that we have less and consequently we should use less; or we may try to create what we lack; or we may finally regard the problem as a worldwide crisis. In a desperate attempt at impartiality the author fabricates economic arguments that the cost will be approximately the same whichever course we choose.

14 A MODEL FOR THE FUTURE 288

In which the reader will find that adolescence, the present stage of energy's development, is a time when advice and predictions are amply available, though not always requested or helpful. An attempt is made to see how promptly troubles can be resolved, and what might happen in the remainder of the century in the United States and in the rest of the world.

EPILOGUE 310

INDEX 313

PREFACE

The first draft of this book was based on lectures given by the author in 1975: the Harvey Prize Lectures, Technion (Israel Institute of Technology), in Haifa, in June–July, 1975; and Distinguished Visiting Lectures of the Faculty of Natural Sciences and Mathematics of the State University of New York at Buffalo, in September, 1975. One chapter is based on a lecture delivered at Acadia University in Wolfville, Nova Scotia.

The lectures in Israel were delivered in accordance with the rule that a recipient of the Harvey Prize travel and lecture in Israel for a period of two weeks. Those two weeks remain in my memory as one of the most wonderful periods in my life.

Among the many people whose advice I sought and employed, I should single out Arthur Kantrowitz and Gary Higgins. Truly valuable suggestions have come, however, from a couple of hundred of my good friends. Of these, Petr Beckmann should be mentioned. I would have attempted to imitate his wide-ranging, clear, and pungent style had I not felt that the attempt would be hopeless. I wish to thank my friend George

Bing, scientist and resident artist at Lawrence Livermore Laboratory, for a number of ingenious illustrations.

Lawrence Livermore Laboratory–University of California, and Hoover Institution, Stanford, California, between which two locations I divide my time, have made, in a variety of ways, real contributions to this book.

It was a pleasure to work with each of the three persons who helped in the preparation of the manuscript. I am grateful to Ann Fogle and Susan Cassell for their work on its early stages. Special thanks go to my assistant Georgia Stoll, who, among other difficult jobs, managed to convert my original version—in a language purporting to be English but more closely related to Hungarian—into its present form.

Ten years ago the United States Information Service asked me to deliver some lectures on the coming energy crisis. I enjoyed doing so, and have continued the practice to this time, particularly since our national Information Service is almost unique among the institutions I know in placing not a shadow of restriction on the opinions, sometimes radical opinions, of the lecturers with whom they cooperate. My own interest in the energy problem was continually stimulated and influenced by these visits in other countries.

My experience was enlarged through collaboration with TRW, Inc., and the MITRE Corporation. Both of them are deeply interested in energy research. In particular, the MITRE Corporation sponsored my visit to Indonesia in 1975, for two purposes: to educate and to be educated myself.

In a peculiar way I am indebted to the antinuclear forces that staged the great nuclear debate of 1976 in California. Confucius said that if you walk between a wise man and a foolish one, you should learn from both.

On the side of wisdom I had the particular advantage of working in 1974–1975 with the Commission on Critical Choices for Americans. Discussions within this group stimulated the energy report I wrote four years ago, and which I must now acknowledge as being overly optimistic. Many of my viewpoints were modified and clarified in that stimulating group. My greatest thanks of all go to Nelson A. Rockefeller, to whose memory this book is dedicated.

February 1979 EDWARD TELLER

INTRODUCTION

This book has one purpose: to give information on a subject that is controversial and complex. I do not claim to be objective. I have arrived at conclusions and my conclusions influence my presentation.

I have a bias. In fact, I have more than one bias. It seems to me proper to disclose at the outset my main conclusions and my principal bias.

A real energy problem exists in the United States, in the industrialized democracies, and probably even more critically in the Third World. If we do not solve the problem its consequences will be severe.

The problem can be solved and the United States can contribute significantly to its solution. This is true in part because of the great natural resources of America; but even more important is America's old tradition for practical innovation—indeed a tradition of breaking whatever is traditional in order to replace it with better tradition. Perhaps the most significant American invention is the permanent, nonviolent revolution. I am strongly biased in favor of nonviolent revolution.

No single prescription exists for a solution to the energy problem. Energy conservation is not enough. Petroleum is not enough. Coal is not enough. Nuclear energy is not enough. Solar energy and geothermal energy are not enough. New ideas and developments will not be enough by themselves. Only the proper combination of all of these will suffice.

The present plight of energy reminds me of the difficulties of an adolescent. He is in a state where the problems of growth and no-growth exist. He is apt to turn to radical solutions; he is apt to exaggerate; he is apt to despair. Yet his problems are not insoluble. A great part of the solution lies in balance.

Because this book originated in lectures, the material is diverse and in some instances there is a lack of cohesiveness among chapters. The subject matter has many facets, resulting again in a certain lack of cohesion among individual portions that are discussed. Therefore I shall attempt to give here an overview of the material. This may help the reader to follow a thread which runs, albeit tenuously, through the book.

The first three chapters describe the cradle of energy, or rather its turbulent origins. This portion is not intended to contribute to any solution. It is, I hope, free from bias. Though it does not contribute to a solution of the problem, I hope it does contribute to the reader's comprehension of it.

A really practical reader may start with the description of the conventional methods for dealing with the energy problem. In Chapter 4, energy begins to grow up. It also gets into trouble. The stage is set in that chapter for problems that must be solved.

Chapter 5 concerns conservation. It discusses an incontrovertible need, an obvious necessity for the adolescent energy: no more waste! So much more easily said than done. Still, an attempt is made to go into detail. The question is not whether to conserve; the question is how to conserve.

The next two chapters, 6 and 7, deal with the classical fossil fuels. Three related subjects run through these chapters: conventional exploitation of these fuels; government regulations, which become unavoidable in times of trouble but which can be overdone; and new methods that could increase the availability and usability of fuels, with minimal disadvantages.

Of course, conventional energy sources by themselves will not suffice. So we turn to unconventional sources. The next four chapters, 8 through 11, are my favorites. They deal with nuclear energy. I worked on this source for half of my life; therefore, I cannot help but give a disproportionate amount of space to it.

This does not mean that I advocate all forms of nuclear energy. With regard to fission energy I was a doubter for many years. Now I believe that fission reactors can make an important contribution in the near future. Nonetheless I do not advocate a nuclear *option*. I believe that nuclear energy is an important component of the solution.

The role of fusion energy is very different. The most exciting research is in this field. We get a glimpse of how the future might look, and what we see is fantastic. But in the near term, to help with the present growing pains our adolescent suffers, fusion energy offers limited hope.

Finally, in Chapter 12, I speak of natural energy sources, discussing specifically solar energy, geothermal energy, and a few other proposals. Chapter 12 deals with quite a few old, even some ancient, energy sources that have recently become popular and that could be really useful when adapted to the needs of the modern world. At the same time, these are the fields where experience, from the distant or recent past, offers minimal information. Prospects for the future, however, are good.

Here the discussion of specifics concludes. What remains for the last two chapters is the outlook. How does our adolescent get out of his immediate difficulties? Will he grow up, and if so how? For human society, and for a specific problem such as that of energy, there may never be such a thing as maturity. It is natural, therefore, that I do not offer a plan nor even a prediction. I make qualitative statements, I give numbers, for one purpose only: for comparison with other points of view.

I have been told never to make a prediction, particularly not about the future. I come dangerously close to making predictions in these final chapters, however. I do not doubt that the future will prove me wrong; the question is only in which particular way I shall be proved wrong.

My purpose is not to be right.

My purpose is not to convince anybody.

My purpose is to give information from my particular viewpoint. And my hope is that this information will be interesting and as close to truth as I can make it.

A few remarks about time scales may help put matters into perspective:

☐ The energy crisis may have disastrous consequences in five years.
☐ We can make predictions and plan technological innovations with some confidence for fifty years.
☐ America was discovered five hundred years ago. When trying to discuss the next five hundred years, we may remember the changes in the last half-millenium.
☐ The next ice age may well arrive in five thousand years, unless technology finds a way to prevent it.
☐ The coming ice age, in turn, may run its course in fifty thousand years. Will any of our descendants still be alive?
☐ Fire was discovered half a million years ago. That was when the human race began.

There seems little reason to list time periods farther in the past except, perhaps, to add that our sun started to shine on a newly formed earth five billion years ago. But for our starting point we shall go back to an even more distant and uncertain date.

PART I

ENERGY'S ORIGINS AND SOURCES

COSMIC ORIGINS

In which are described the guesses that scientists make about the universe, the galaxies, the stars, and stellar explosions. All known sources of energy are derived from these cosmic origins.

In our many discussions of contemporary world problems there is one inevitably recurring question: How will it all end?

It seems curious that this question should so occupy us. We have problems enough for the next year or the next decade, yet temporary solutions do not seem to satisfy. War must be abolished for all time; pollution must be eliminated for good; energy must flow from renewable sources so that shortages will never again occur. Such demands, of course, reflect attitudes of the moment. They reflect the habits of thought of many people.

This preoccupation with end results may even have something to do with an imitation of science where, indeed, problems often are solved in a manner so striking as to appear final. In the practical world, however, ultimate solutions are usually not available.

Seeking ultimate solutions nevertheless has some obvious advantages, and one of them appeals to me. When you seek ultimate solutions you begin to think in broad terms, and you may be able to put some distance between your thoughts and the problems calling for solution.

I shall use this style and consider energy at first in the broadest possible terms. How did our energy sources originate? What will happen to them in the end?

One should not believe that the tentative answers we can give to these questions have of necessity a bearing on the practical problems with which we struggle. But a positive consequence of the energy shortage is that it stimulates curiosity. Curiosity is one of the most valuable properties we inherited from our simian ancestors. It often leads to ultimate enlightenment, and even to useful applications.

It is by no means obvious that the world had a beginning. Neither is it certain that it will end. The belief is popular, however, among laymen, theologians, and scientists that the universe had a "first day," or an initial microsecond, and that it will end, wither away, or "run down" in some manner. The most widely accepted hypothesis among scientists at this time is that the universe started with a "big bang" some fifteen billion years ago. Also, it is frequently assumed that all processes will run their course, ending, after many more billions of years, in an uninteresting equilibrium which we might envision as a universe burnt out and therefore dead. While making these statements I am reminded that ideas about the beginning and the end of the universe were very different two hundred years ago. More particularly, our present concepts of the origin of the universe were suggested a mere half century ago.

Energy plays an important part in our ideas concerning both the beginning and the end. In the beginning everything was exceedingly hot. At the end the universe is apt to be cold and dark. The beginning and the end are far away in time, however, and our ideas may be the result of limited imagination; rather than dreary and cold, the end may be hot as blazes.

It is imagined that at the beginning matter was present at an enormous density, far more than a billion times the density of familiar materials. It is also imagined that temperatures exceeded at least a billion times those which we encounter on earth. We know so little about the true beginning of the universe that, at this time, it makes no sense to say how small a volume any portion of the exceedingly dense original matter occupied. But speculations do go back to a state where all the matter in the universe that we can see (that is, from which at present we can obtain light signals) was contained in a sphere comparable in size to our solar system. According to present theories, the universe may collapse after many billions of years into its exceedingly hot and dense original state.

According to some theories, the universe was and is finite. Describing it is not unlike describing the surface of the earth, which, of course, is not infinite, but is without boundaries. Columbus never had a chance to sail off the edge of the world. For the two-dimensional surface of the earth we can "see" this. Whoever looks at a globe has a vivid picture of the earth's curvature. It is harder for us to visualize that the universe is "curved." We would have to have the capacity for a nonexistent four-dimensional perception in order to visualize a finite three-dimensional space with no limits. But mathematicians can formulate and quantify even those things which neither they nor anyone else can perceive.[1]

It is easiest to assume, and therefore it is assumed, that the basic material was that of the simplest element, hydrogen. You may think of the universe as consisting of many hydrogen nuclei, or protons carrying positive charges, and an equal number of negative charges, much lighter, which are called electrons. How many? The conventional, uncertain answer for the universe visible at present is a little less than 10^{80}, which is 10×10 repeated eighty times.

At the very high temperatures we believe existed, however, the components of the hydrogen atom, protons and electrons, were by no means

---

[1] *Flatland,* a beautiful fairy tale on the subject of how many dimensions there are, was written in the last century by a mathematician, Edwin Abbot. Application of these ideas to the three-dimensional physical world and to the universe was first made by Einstein.

the only occupants of space. As electrons whiz past protons they must emit light. We now know that at high energies many other radiations and particles are emitted, many of which have been discovered in the past few decades, and many of which are ephemeral. They disintegrate in a time shorter than what the reader would consider instantaneous,[2] but at high temperatures they are formed again and again.

One absurdity usually not discussed even among professionals is that in spite of temperatures greatly exceeding those of Dante's inferno, and in spite of the enormous density of energy present, none of the energy would have been available for the purpose of performing useful functions. The random disorganized motion characteristic of heat cannot be used to perform work or most of the other functions associated with the common conception of energy. We can think of an inferno compared to which fire is cool, where no pitchforks, let alone devils, could have existed. In order to perform a useful function, energy must be present in a more orderly form. This is the case when particles in a certain volume move predominately in the same direction. This motion can perform useful work. Such useful work is produced by engines running on fuels that generate heat. The heat causes differences in temperature, which, in turn, bring about orderly motion (for instance, by a force acting on the piston of an automobile engine).

In the original imagined state of chaos, no such differences of temperature are assumed. How, then, could the potential for useful work ever arise?[3] The answer is that the universe was not and is not static. The early stage is actually described as exploding. The explosion was accompanied by rapid cooling, in the course of which some processes did not manage to get into equilibrium. As a result, chunks of energy-rich materials, fuels, were left behind. One of these fuels is heavy hydrogen, also called deuterium, now considered a potential, practically inexhaustible fuel. It might be in general use in the next century.[4]

Some of the particles that probably existed in vast numbers during those early hot stages were an excess number of electrons, a corresponding

---

[2]This seemingly absurd statement means that the longest of these times is two-millionths of a second.

[3]There is a branch of exact science, called thermodynamics, which states that the potential for useful work diminishes with the passage of time in any clearly circumscribed static system.

[4]The use of this fuel, deuterium, will be discussed in Chapter 11.

number of positively charged electrons called positrons, and a great density of peculiar particles called neutrinos.[5] Neutrinos are very hard to detect and live forever. Even today they are all around us and continue to pass through us with the velocity of light. We don't notice them because their interaction with other forms of matter is incredibly weak.[6] Their significance is that when an electron and a neutrino hit a proton at the same time they can disappear into the proton and change it into a neutron. The proton and neutron are similar particles except that the neutron carries no charge. The process that transforms one into the other does not occur at a high frequency unless a superabundance of electrons, positrons, and neutrinos exists, as was the case in the original hot chaos.

As the universe expanded and cooled, the density of electrons, positrons, and neutrinos diminished rapidly. Furthermore, as the temperature dropped, the ratio of protons and neutrons shifted in favor of the more stable protons. Finally the rapid transformation came to a virtual halt, leaving about 20 percent neutrons and 80 percent protons, according to present estimates. The remaining neutrons should have changed into protons in less than an hour; but there was no time for that. Something else happened before this transformation could occur.

Neutrons and protons are the constituents of all atomic nuclei. They attract each other and, at present, form within the atom an extremely dense central body with a radius less than 1/10,000 that of a normal atom.[7] Present speculation is that before all neutrons were transformed back into protons, some of them combined with protons and formed *deu-*

---

[5]"Little Italian Neutrons," first discussed by the great Italian physicist Enrico Fermi, who himself was small only in physical appearance.

[6]Interactions lead to deviations from movement along straight lines; they may also lead to depositing of energy. Today it is supposed that neutrinos are generated near the center of the sun, the region from which all solar energy originates. Neutrinos carry the greater portion of solar energy. They emerge from the sun without interrupting their travel. They then cross the whole of the earth without leaving behind more than one-billionth of their energy.

[7]In the early state of the universe which we are envisioning, atoms could not exist because there was not enough room for these relatively big structures in the superdense state of matter. It is one of the peculiarities of the theory of the atom and its parts that if you tear atoms apart into protons, neutrons, and electrons, and make everything move at very high velocity, less room is required.

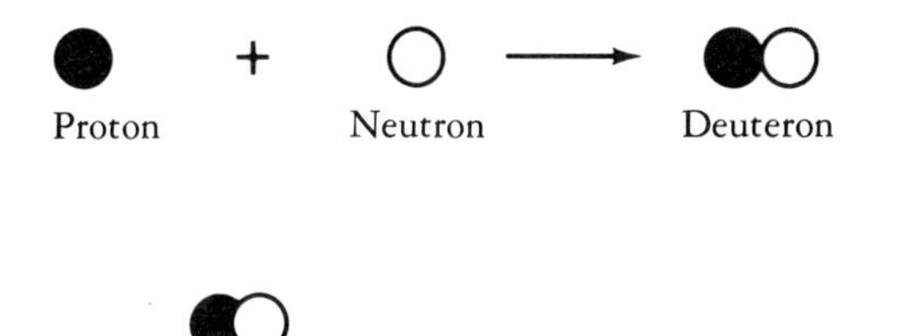

FIGURE 1-1 *The formation of a deuteron.*

FIGURE 1-2 *The formation of a helium nucleus.*

terons, [8] or the nuclei of heavy hydrogen. A deuteron is symbolized in Figure 1-1, where the black circle represents the proton and the white one the neutron.

Deuterons were abundant. But they are weakly bound and they undergo further reactions. They met each other and produced nuclei containing two neutrons and one proton (with one proton left over) and other nuclei containing two protons and one neutron (with one neutron left over). These nuclei reacted further with the deuterons and formed the stable alpha particles that are the nuclei of common helium atoms.[9] These consist of two protons and two neutrons, and may be pictured as two black and two white circles. Needless to say, these pictures are strictly symbolic; neither they nor any similar pictures should be taken to represent reality in any sense.

The way these early nuclei were formed has been calculated. In the process approximately as many helium nuclei were formed as are now found in the universe. Indeed, hydrogen is the most abundant element in the stars, and helium is the second. Some heavy hydrogen (deuterium) was left behind in an abundance calculated at approximately two parts in one hundred thousand. This calculation corresponds to the best evaluation of the current cosmic abundance of heavy hydrogen. Calculation of such abundances actually provides evidence—perhaps the only quantitative evidence—that the big bang idea conforms to reality. But we should not be overly impressed. A few pieces of evidence don't prove a theory.

Here on earth the available heavy hydrogen is more abundant than it is on the average in the universe. Heavy hydrogen constitutes one part in fifty-five hundred light hydrogens. There are several ways to explain why we have such a big share of this high energy fuel.

---

[8]If you know Greek the names are easy to remember: the first particles are protons, the second are deuterons.

[9]These reactions are discussed further in Chapter 11.

We can summarize. In the beginning there was matter and heat but no usable energy. In the first half hour of creation, according to our story, the first complex nuclei were formed; among them was heavy hydrogen that could deliver usable energy for earth's future inhabitants.

Having succeeded (we hope) in determining how deuterium and some other light nuclei were produced, we find that the temperature dropped sufficiently to slow down further nuclear reactions to an imperceptible rate. All atomic nuclei carry positive charges. They repel each other. Unless the temperature is exceedingly high and all particles are in violent motion, nuclei do not get close enough to each other to exchange protons and neutrons and to produce bigger aggregations of these two building blocks of atomic nuclei.

We cannot properly speak of atoms during this formation period. The density was still so great that electrons were not individually attached to any atoms, and their temperature was so high that they could not be held in quiescent, stable atomic structures. A few million years later, when the density became low enough and matter had cooled sufficiently, most electrons got attached to nuclei, and atoms of the kind we know—mostly hydrogen atoms—were formed.

The next stage in the evolution of the universe, many millions of years later, was not further formation of fuel or change in atomic structure, but the creation of bigger structures: stars and galaxies. Of these, the galaxies were formed first. A galaxy is actually an assembly of a huge amount of matter sufficient to supply the substance for a hundred billion (that is 10^{11}) stars. The universe in turn contains billions of galaxies.

The original uniform distribution of matter in the chaos became more organized, and approached the state that the Greeks called "cosmos," or beautiful order. That this could happen was, again, due to expansion and cooling.

Expansion and cooling have actually two opposite consequences. One diminishes the effect of gravitation; the other indirectly enhances this effect. Consider a finite but very big fraction of matter in the universe. If this matter expands so that it occupies a space twice its initial diameter, the potential energy, causing gravitational forces that try to pull the matter together, decreases to one-half its original value.

At the same time, expansion decreases the disorganized velocities within the matter. One may imagine a perfectly elastic ping-pong ball bouncing around in a room in which the walls are receding. Due to the recession of the walls, on each collision the ball bounces back with less velocity. By the time the room is twice as big in diameter, the ping-pong ball's velocity will be reduced to one-half of what it used to be. The same holds for particles in the expanding universe. Half the velocity, however, corresponds to one-fourth the kinetic energy, as should be obvious to automobile drivers, who realize that a collision at reduced speed is much less dangerous than one at high speed.

The two-fold expansion, in linear dimension, has reduced internal gravitational energy to one-half but kinetic energy to one-fourth. Therefore, gravitational energy has a better chance to prevail over kinetic energy. This stops the expansion of a portion of material. This material stays permanently together, and an island universe, or galaxy, is formed. Galaxies, in turn, move apart, and the expansion of the universe as a whole continues.

How do these galaxies come about, and what is the natural size of a galaxy? These are difficult questions, and we do not have completely satisfactory answers. In answering, we must not only consider the disorderly heat motion, but must also remember that all parts of the universe are engaged in a uniform expansion. We must take into account the effects of radiation, which was extremely important in early stages but later played a decreasing role.

Gravitational attraction was the main reason that at some stage, probably many millions of years after the big bang, the matter of the universe was divided into galaxies. The galaxies were of different sizes and shapes—spherical, elliptical, and frequently spiral, as our galaxy is. It is clear that many galaxies rotate, and this rotation had to be present at the time these enormous chunks of matter were separated.

A galaxy contains enormous amounts of gravitational energy and random kinetic energy; a rotating galaxy also has less, but not much less, rotational energy. I cannot conceive that these energies could be used by us or by some trans-Martian who might be in some distant part of the universe, with one possible exception.

In the center of some galaxies there is evidence of violent processes, reminiscent of quasars, unexplained distant energy-radiating objects. That these violent and turbulent regions might contribute to the "ecology" of the universe is a thought beyond science fiction. (We shall return to this idea toward the end of this chapter.)

The concept of the expanding universe is paradoxical. Physicists talk about conservation of energy: energy can never be created, can never be lost. Yet we continue to speak of the cooling of matter in the uni-

FIGURE 1-3 *Andromeda, our sister galaxy, approximately two million light years away, is a spiral galaxy like our own. This assembly of approximately a hundred billion stars has a diameter of about one hundred thousand light years. In an analogous picture taken from Andromeda our sun would not be visible, but should be located in the outer part of the structure. (Lick Observatory photograph.)*

verse, which corresponds to a decrease in heat energy. If compressed gas pushes a piston and thereby expands, its heat energy is transformed into the kinetic energy of the piston. One could reverse the process and transform the kinetic energy of the piston back into compression and heat. In the case of the expanding universe, this transformation may or may not take place. In any case, the concept of conservation of energy is not so easily formulated when we speak of the universe as a whole.

Actually, an argument can be advanced that the total energy in the universe is conserved, when energy is defined from the point of view of one observer located in one of the galaxies. But he will see that most of the energy is present in the form of kinetic energy in very distant galaxies which are receding at very high velocities. He will find that this energy is unavailable to him. Conservation of energy for the universe as a whole, therefore, has little practical meaning.

Another point is even more remarkable. We have seen that starting from chaos we are approaching some sort of ordered cosmos. In the usual physical processes, the opposite is happening. One basic natural law is that disorder is always increasing. How can this law be violated?

The reason is that the volume of the universe continues to increase. The usual measure of disorder would become ever greater if material were filling the increasing volume of the universe in a uniform way. What happens is that the disorder is not increasing as greatly as could be the case, considering the availability of greater volumes.[10] We can use an all too human comparison. Consider an ideally disorderly secretary who manages to keep her filing system in tolerable and ever increasing order by utilizing, at an ever increasing rate, new filing cabinets. Another way to express this concept is to state that disorder in the universe is not only great, but ever increasing. Fortunately, the disorder never becomes quite as great as it could, and it is therefore possible to use the incompleteness of disorder to the end of performing orderly changes, which in physics we call work.[11]

How does this account of universal disorder agree with our ideas about the end of the universe, when it is imagined that everything assumes

---

[10] It can be shown that in the absence of interaction among particles, a completely disorderly universe would remain completely disorderly. This does not hold, however, when particles interact by gravitation or in other ways.

[11] We shall return to this discussion in Chapter 5, where we shall find that what is called energy conservation really means the prevention of unnecessary disorder.

uniform temperatures, and there is no more energy for work or for orderly change? Our present picture of the beginning does not agree with the usual picture of the end. We have started from uniform temperatures, yet, within a few minutes of the big bang, we are left with actual useful fuel. In less than a million years more order was generated as the galaxies emerged.

The conclusion that complete equilibrium and equal temperatures lead to an unchanging "dead" situation is true for an enclosure which itself does not change. If we let that enclosure expand, work will be done on the expanding enclosure—on a piston, for instance, which, in a machine, constitutes one wall of the enclosure. If expansion is too rapid, equilibrium inside the enclosure will be disturbed and correspondingly less work will be done on the piston. Instead we get the chance of more work inside the enclosure. This additional work, in the case of the universe, has been stored up for potential use fifteen billion years later.

If we consider the universe as a whole, work done by the expanding universe on the outside makes no sense. There is no outside. It follows that a conclusion about the universe running down at some time is not convincing. As long as there is expansion, there is a chance of creating more opportunity for order that may be utilized.

In less than the first hour of creation a considerable amount of fuel was produced and conserved for later exploitation. What we have discussed has no direct bearing on practical energy supplies, but is only intended to relate the evolution and changes in the universe. Now we can return to truly important energy sources, the stars.

We know that within the galaxies the formation of stars has proceeded, and is proceeding still. Like galaxies, the stars form by contraction. They are still massive, but small compared to a galaxy. Inside the stars, high temperatures are produced by gravitational contraction. At these high temperatures, nuclei can approach each other, and bigger nuclei can be built. In general terms, this process is the source of the radiant energy of the stars, among them our sun.

Before we discuss solar energy, however, we should consider other spectacular events that preceded the formation of our solar system. We have an historical record of the first observation of such an event. In the summer of A.D. 1054, during the reign of China's Emperor Jen Tsung of the Sung dynasty, in the Chih-Ho period, the Chief Calendrical Computer reported the appearance of a "guest star" near the configuration of Aldebaran. The star was so bright as to be visible in the daytime. The report seems to have been submitted with considerable trepidation because the

FIGURE 1-4 *A beautiful spiral galaxy called the Whirlpool, seen face on, with its smaller companion galaxy. It is about thirty million light years from earth. The rotation is obvious. (Lick Observatory photograph.)*

predecessor of the imperial computer of stellar events lost his head for failing to predict a solar eclipse. The Chief Calendrical Computer, whom we may call the Imperial Astronomer, promptly wrote a memorandum congratulating the emperor on the star which, he said, meant great good fortune to the emperor and to the House of Sung.

Happily for the astronomer, the emperor seems to have been less interested in a single star than in a solar eclipse. But his lack of interest was hardly justified; that star was a *supernova,* a stellar explosion that shone for a few weeks with the brilliance of a whole galaxy. In order to understand some of the fuels now available to our nuclear industry, we should discuss what went on inside that Chinese supernova, and other supernovae that occurred before and have been observed since.

Our sun is one of myriad small stars. It consumes its nuclear fuel in the course of many billions of years. During this burning process its properties change in a slow and undramatic fashion.

Bigger stars are much brighter. They exhaust their fuel, and then are apt to collapse. Their centers become dense and very hot. Gravitational forces may become so powerful that nothing, not even a ray of light, can escape. The result is a *black hole,* the like of which was never seen, but which probably is present in great numbers in every galaxy.

I do not like the name "black hole." I prefer the word "Boojum," introduced by the mathematician Lewis Carroll in his epic poem "The Hunting of the Snark." He is better known for his research on Alice in Wonderland. According to Carroll, the consequences of running into a Boojum is that "you'll never be met with again," a most concise description of a black hole.

But what we want to understand is the opposite of a stellar collapse. The supernova is a star that emits in a few weeks more radiation than our sun has produced in its five billion years of existence. The reader may notice that our sun, according to what is said here, is not as old as the universe. We shall return to this point later.

The explanation proposed for such stellar explosions, which are of course much smaller than the big bang, is that a stellar collapse, caused by gravitation, is followed by the explosion of part of the mass of the star. The great heat just outside the region of most violent collapse might suddently produce reactions between atomic nuclei, which release a lot of energy. The result is that those parts of the star that are sufficiently far from its center will be blown outward.

It is more probable that the explosion, which follows the implosion, has a different cause. Near the hot dense center of the collapsing star, neutrinos are generated in great abundance. (We encountered them in the first few minutes after the big bang.) These neutrinos penetrate material with ease—except when the material is extremely dense. They transfer outward-directed motion to layers surrounding the dense core, and thus initiate the explosion.

It may be worthwhile to reiterate this last sequence of hypothetical events. In the collapse, gravitational energy is transformed into heat, and neutrinos are emitted. Portions of the collapsing star are situated at the outer edge of its extremely dense core and are not as strongly attracted to the hot, heavy center. The neutrinos carry energy into those portions. The outer layers then are blown away in a great explosion.

Part of this hypothesis is verified by observation. The Chinese Imperial Astronomer pinpointed the place in heaven where his bright, short-lived star burst into sudden brilliance. If we look there we see no bright star. What we see is the Crab Nebula, an irregular mass of material blown away from the supernova. Today, more than nine hundred years later, this mass is still in violent motion.

There are problems with the explanation of supernova explosions caused by neutrinos. One is that some calculations do not lead to sufficiently explosive effects. Most of the material continues to fall inward and is swallowed up by a black hole. Slight changes in the assumption on which the calculations are based seem to substantiate greater explosions, in accordance with observations.

The other difficulty is the black hole itself. At the place of the Chinese supernova, in the middle of the Crab Nebula, we don't see a black hole. (In fact if it were there a black hole would be invisible; its existence can be determined only by study of its space environment.) But we do see a rapidly rotating pulsar, a small, heavy, dense star that performs about thirty revolutions per second. This is the dense remnant of the supernova. Its rotation seems significant; it is one of the conditions missing from the neutrino explanation described above.

If a heavy star rotates rapidly it can never suffer a fatal collapse. Contraction causes rotation to speed up and centrifugal force prevents the collapse.

A nonrotating heavy star may collapse into a black hole and blow off a small amount of material, in which case relatively little energy would be left for an observable explosion.

A heavy star rotating at a moderate rate, normal among stars, will collapse; but the final phase of the collapse, the formation of the black hole, might be prevented. Instead a pulsar will be formed, and more material may be blown off in an event corresponding to the observed supernovae.

It should be emphasized that as yet no calculation exists to prove the hypothesis I have described. Indeed, making such a calculation would be extraordinarily difficult. It appears likely that the mistake in early, un-

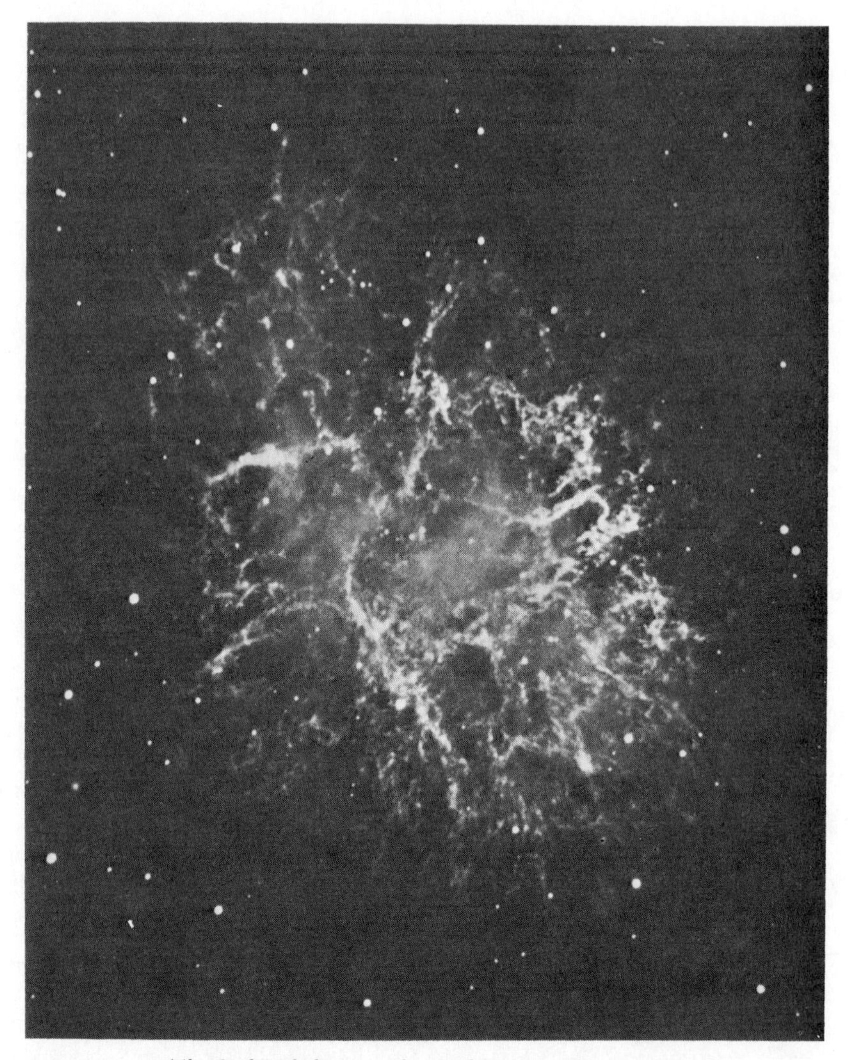

FIGURE 1-5 *The Crab Nebula is six thousand light years from earth. The irregular filaments are ejecta from the eleventh-century "Chinese" supernova, which follow tangled magnetic fields. Even the immobile picture suggests the fact of violent motion. (Lick Observatory photograph.)*

successful calculations was due to lack of detailed knowledge about the state of nuclei at high temperatures.

Whether or not details of the explanation are correct, it is clear that supernovae leave a lot of debris behind. In that debris lies the importance of supernovae: our sun and its planets have been formed from material that included the debris of one or more supernovae.

We have seen that one nuclear fuel, heavy hydrogen, may have been formed half an hour after the big bang. When and where did our other principal nuclear fuel, uranium, originate? It should be mentioned that, in general, light elements are the abundant ones. In addition to the lightest and most abundant, hydrogen and helium, all other moderately abundant elements, such as carbon, oxygen, aluminum, silicon, and even iron,

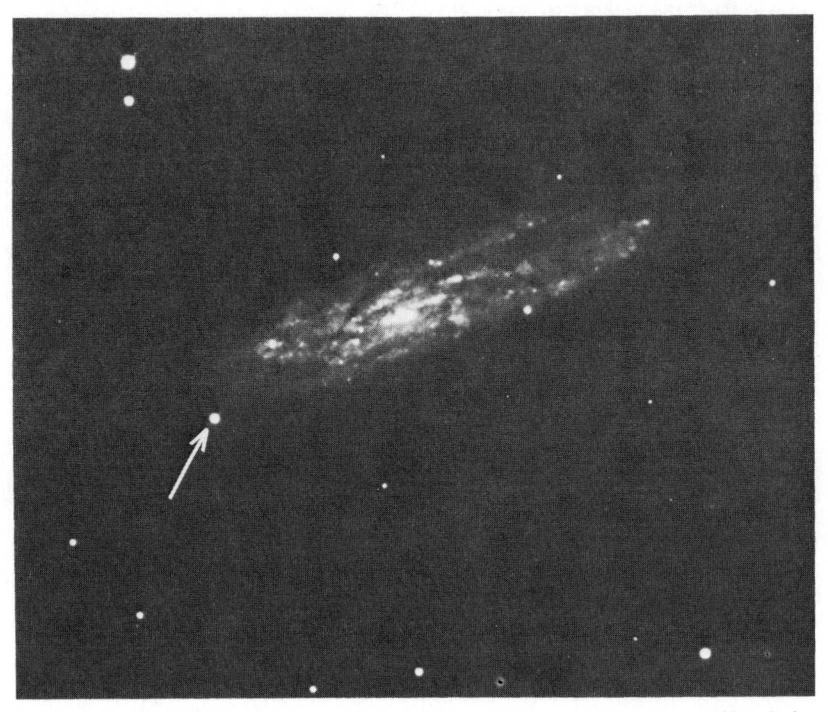

FIGURE 1-6 *This spiral galaxy and its 1961 supernova are about thirty million light years away. The arrow points toward the supernova, which was not there a few weeks before this photograph was taken. The supernova is really as bright as the galaxy of which it is a part. Other distinct stars in the picture are within our own galaxy, relatively close to earth. (Lick Observatory photograph.)*

are relatively light. Beyond iron, elements begin to decline in abundance. In general, the heavier the element, the scarcer it is. It is believed that the heavier elements, including uranium and another similar nuclear fuel, thorium, are formed near the core of collapsing stars.

In this dense core electrons are pressed into protons to form neutrons. Electrons are normally present in outer portions of atoms, while nuclei are composed of protons and neutrons. The dense pulsar is believed to be a neutron star of superdensity, more than a trillion times the density of gold, a density actually comparable to that of an atomic nucleus.

As the material surrounding the core is blown away, we may imagine that chunks of nuclear material consisting mainly of neutrons break loose. This breaking away may be the origin of the heavier elements. Alternatively, heavy nuclei may grow from lighter nuclei with the addition of neutrons. Neutrons are not repelled by the positively charged nuclei as protons are. In either case, the nuclei formed are richer in neutrons than those found in nature. But we know that neutrons can change into protons inside a nucleus, and in the transformation an electron and a neutrino are emitted. The heavy nuclei we find in nature result.

Much of what has been said is speculation, modified and influenced by comparison with known facts. But one conclusion is certain: our solar system and the materials from which it has condensed is, in good approximation, five billion years old.

How do we know this? Radioactivity is a spontaneous process, and it is going forward with perfect regularity, incomparably more accurate than clockwork. Radioactive "clocks" are imbedded in the rocks of earth, in meteorites dropping from the skies, and in samples brought from the moon by the astronauts. Various samples have various ages. But the maximum age on earth, among meteorites, and in the material from the moon, is always near five billion years.

Uranium on earth contains mainly two kinds of nuclei, or *isotopes*[12] with two different kinds of nuclei: relatively inactive uranium 238 and the rare and valuable uranium 235. The former is present in the amount of 99.3 percent and contains 92 protons and 146 neutrons; the latter contains 92 protons and 143 neutrons. The two kinds have similar chemical

[12]Isotopes are atoms practically identical to each other in their chemical behavior and peripheral electronic structure, but have different nuclei.

properties and are difficult to separate from each other. Both are radioactive. Half of the abundant uranium 238 disintegrates in five billion years, half of uranium 235 in seven hundred million years. In the future, uranium 235 will become even less abundant relative to uranium 238 than it is at present. It follows that in the past uranium 235 was relatively more abundant. Five billion years ago uranium 235 was nearly as abundant as uranium 238. Six billion years ago uranium 235 should have been more abundant than uranium 238. Ten billion years ago uranium 235 would have been approximately one hundred times more abundant than uranium 238.

We have no other example where a nucleus with an even number of protons (92 in uranium) prefers to have an odd number of neutrons (143 as in uranium 235) rather than an even number (146 as in uranium 238). The nucleus with the even number is always generated in greater abundance. And nuclear theory gives plausible reasons for this empirical rule. It follows that our uranium could not have been produced much more than five billion years ago.

The age of the solar system is five billion years. Our sun is not one of the original stars in the universe. The material that we find here on earth originated less than a billion years before the formation of the solar system. It probably has been "cooked" in the hot furnace of a supernova.

This completes the account of the cosmic origin of our nuclear fuels. The light ones were made in the big bang;[13] the heavy ones in one or more supernovae, which scattered the material that was to condense and form the sun and the planets.

This does not complete the story of the cosmic origin of energy. For us the most important source of energy is and will be our sun. Solar energy, in turn, is produced in the hot center of the sun. Here hydrogen nuclei, which are protons, collide. In most of these collisions nothing happens. But once in a very long while two colliding protons produce a heavy hydrogen, a positive electron (positron), and a neutrino. A chain of reactions follows in which the heavy hydrogen is burned up and vastly more energy is liberated.

Energy released near the center of the sun takes almost a million years

---

[13]This statement needs one correction. One important light material, lithium 6, was probably not made in the big bang.

to reach the solar surface, and is radiated away into space and onto earth. What happens to it on earth, how it is used, and how it is stored we shall find out in Chapter 3.

There is one other source of energy which I want to include in this chapter: the heat energy of earth. Perhaps what is found here on earth should not be called "cosmic." But something associated with the earth as a whole is similar to other cosmic events.

When the earth was formed, gravitational energy was transformed into heat. That heat is still there, near the earth's center, where we find a great amount of molten iron.

Additional heat is released by radioactive sources within the earth mantle. It is a curious fact that most radioactivity in the earth, including most uranium, is found within a few miles of the surface of our planet. Why? One obvious reason is that radioactive materials release heat, becoming liquid and somewhat less dense. One may say that they melt their way to the surface.

Another and perhaps more important reason is that radioactive substances are relatively rare. It also happens that they combine rather easily with oxygen. Oxides cannot dissolve in the liquid iron near the earth's center, and they do not fit into the solid lattice structure of materials in the earth's mantle. So they are left over as "slag" near the surface.

It is amusing to consider why gold is particularly rare. It cannot combine with oxygen, and the pure metal dissolves in the molten iron deep within the earth. In this way gold has become inaccessible, and therefore valuable. Lead, on the other hand, and the nuclear fuels uranium and thorium, do form oxides, and are found in greater abundance in the accessible crust of our planet.

There is a lot of hydrogen in the universe, but relatively little is found on our planet. It was too light to be attracted by the material accumulating to form the earth. What there is of it exists in the form of water. Water was spewed out in the course of billions of years from volcanoes. But the heavy hydrogen content of the water on earth is approximately ten times greater than the heavy hydrogen found in the universe. And we cannot complain of a scarcity of heavy hydrogen. From the standpoint of energy, there is enough of it.

Should we consider that the fuels we have discussed are renewable? In a strict sense, nothing is renewable. Even the sun, though its lifetime is measured in billions of years, will consume its fuel and shrink into a *white dwarf,* a dense (but not extremely dense) star of slight luminosity.

FIGURE 1-7 *Fifty million light years away, Messier 87 is a nonrotating galaxy with an unusually dense and active center. The center has emitted the jet (a strong radio source) seen in the photograph. A most violent process must be going on in the center. It may be the growth of a black hole incomparably larger than black holes connected with supernovae. (Lick Observatory photograph.)*

Nuclear fuels—deuterium, uranium, thorium—could last for millions of years if we exploit them in a careful and complete manner. They are not renewable, but from our human perspective they are inexhaustible.

Geothermal energy, the heat energy of the earth, is different. What comes from great depth is a relatively meager source of energy. That part which is stored in the few miles nearest the surface can be used effectively. But this resource is no more abundant than coal. We may exhaust it in a few centuries.

Perhaps there is one lesson that might be drawn from this chapter. Our knowledge of the processes by which energy sources are formed is incomplete, but our list of energy sources may be complete. We have looked far back into the history of the universe. And we probably know enough about fuels to say that no more completely novel sources of energy will become available in practice.[14]

---

[14]A hundred years ago, nuclear energy was not even a dream. At that time the claim that we know all fuels would have been wrong. But at that time a good scientist would have made no such claim. He would have realized that he did not understand the energy source of the sun.

There is, however, an indication of a remote additional energy source. Some galaxies have exceptionally hot, active centers from which we receive signals that are incompletely understood. From such centers emerge nonuniform jets that may well be due to explosions compared to which a supernova would be a firecracker. Will the explanation of this phenomenon, or of similar violent events, ever lead to additional useful energy sources?

The universe is big, and usable energy resources are enormous. Man and his needs are tiny. There is enough energy even on this small planet where most of us seem to be confined for the foreseeable future, a most uncertain time span.

CHAPTER 2

SUN, EARTH, AND LIFE

In which it is shown that our sun is a remarkably constant star. It has scarcely changed in billions of years; it is not likely to change in many more billions. Life has developed on earth during approximately three billion years. A time equal or even longer remains for our descendants.

We have discussed the origin of nuclear fuels. Another energy source is at least as abundant and reliable as nuclear fuels ever will be. It is energy from the sun, which has nuclear origins.

Finding a practical way of using solar energy in an industrial society is a problem as yet unsolved, which will be discussed in Chapter 12. But all life on earth has depended on solar energy. The history of evolution on earth covering a span of billions of years is our actual evidence that solar energy has been steady and unfailing.

Let us consider the scene in which the evolution of life has occurred. First, some general observations. Our sun is one among approximately a hundred billion sister stars belonging to our galactic system. We can see this galactic system on a clear dark night as the Milky Way. It was Galileo's telescope that first drew attention to the fact that the Milky Way is composed of myriads of stars.[1]

The majority of these stars are small, which means roughly that they are comparable to our sun. Some of these small stars are brighter than our sun; the great majority are fainter. There is also a scant number of stars called *giants,* some of them a thousand times or more as bright as our sun. Giants also contain more matter, though not much more. A star that has ten times the mass of our sun is among the brightest giants.

There is no sharp division between stellar dwarfs and giants, but there is one important distinction. Dwarfs husband the nuclear energy source of their radiation; they last for many billions of years. Somewhat arbitrarily I call our sun a dwarf. Giants are spendthrift stars. They consume their fuel in a few million years and then collapse. As described in Chapter 1, the collapse is probably followed immediately by an explosion. Dwarfs have a different fate. After many billions of years they become cinders which are still slightly luminous; they are then called white dwarfs. Their fate is equivalent to a sedate and uneventful funeral. We have good reason to believe that they will not behave as violently as the giants.

What about life? To begin with, it would be desirable if we could define life. An obvious statement is that life is a peculiar phenomenon on earth, with which all of us are familiar. This is an insufficient definition

---

[1]Some objections came from an interpretation of the Bible holding that the stars were created to please man, not to please man equipped with a telescope, and that therefore Galileo must be wrong. Since a telescope is needed to see some of them, such a goal of their creation would not be attainable.

that will not serve when we ask whether life can be found elsewhere in the universe.

Unfortunately, an adequate definition does not exist. My favorite definition is "life is a little matter coupled with an enormous amount of complication." By complication I mean an incredibly intricate and varied set of chemical linkages. We are beginning to understand the structure of a few relatively simple examples of these complicated molecules. There are vastly more kinds of chemical combinations than there are individual differences among species, and far more than there are individual differences among people. Still, within these complications there are persistent resemblances. Altogether we are faced with a pattern; the most elaborate art form appears boringly simple in comparison.

The question we should raise is whether similar complications can be found anywhere else in the universe. Today we have evidence that they are not found on the moon nor on Mars. In our solar system it seems probable that life is confined to the surface of our planet.

But statements of this kind cannot and should not end speculation. One would guess that life is a phenomenon confined to the surface of planets near stars. They need energy from the stars, but not temperatures so high as to cause violent molecular motion, which would destroy all order and all accumulated complication (because complication is hereditary and is accumulated). Also, on the surface of a planet there is a chance to dump energy into space and thus bring about the temperature differences that are the necessary precondition for useful work and a variety of organized motions. However, the fact that I confine "life" to the surface of planets may be merely evidence of my lack of imagination.[2]

That all conditions should be right for the development of life seems improbable in any instance. But the opportunities are almost endless. There are billions of stars, and probably billions of planetary systems, in a galaxy; and there are billions of galaxies in the universe. I like to believe that in some regions of the universe one could find matter vastly more complicated than we are. By comparison we might appear as primitive as a virus is when compared to us. Thus we may not only be dwarfed by the size of galaxies, but also we may have to regard ourselves as insignificant in comparison with other miracles that may have arisen in distant places.

Let me add one more speculation. Life, and probably any complica-

---

[2]See *The Black Cloud,* a science fiction story written by the scientist Fred Hoyle.

tion, will need a long time to develop. Of course we cannot be certain, but if this is true then no life is apt to be found near giant stars. Looking at supernovae we cannot help but wonder, "What if it happened to us, or to living beings like us?" There are two comforting answers. Our sun is too small, so it can't happen here. And where it could happen, stellar evolution will be so rapid that no higher form of life will have had a chance to establish itself.

Let us return from our imaginings to the sun and to earth. We know of fossils that are five hundred million years old. Before that long-ago time there seem to have been no forms of life to leave a clearly visible trace, such as bone or shell or even the imprint of a leaf.

Yet we know that life is much older than those fossils. In fact, it has been around for at least three billion years. In ancient rocks we find spots of partially decayed organic matter such as would be left behind by complicated proteins. The very existence of complication, of chemical linkages, leaves an imprint that is hardly mistakable even though it is on a molecular scale, and is based on nothing more than the close proximity of a variety of typical molecules containing peculiar carbon chains.

Life has been around during more than half the age of the earth. Therefore, during this long period, temperatures cannot have been so low as to freeze all the oceans. It is estimated that all of the earth would have been covered with ice if solar radiation had been less than at present by even 20 or 30 percent. Once the earth is covered by ice it would be very difficult to melt the ice again. Ice reflects solar radiation well, and at the same time wastes energy by radiating it out into space as long-wavelength infrared radiation invisible to us. Indeed, the surface temperature of the earth results from a balance between radiation received from the sun as visible light and radiation emitted from the earth into space as invisible infrared radiation, also, and somewhat improperly, called "heat radiation." We have, therefore, a good indication that the sun has been hot for a long time.

The sun could not have been much hotter than it is today, however. A twofold increase in radiation for an extended period would have evaporated all oceans. The existence of life on earth for billions of years indicates that there could not have been much change in the energy production of the sun for a very extended period. Of course, the sun is the only star for which we have this kind of evidence.

Actually, the situation is complicated. We have evidence about temperature on the earth rather than about radiation from the sun. Just after the earth was formed, gravitational energy released in its formation must have caused considerable heat, but this original geothermal energy remained available for a comparatively brief period. The center of the earth is still at thousands of degrees. But today the energy arriving from that depth is only one ten-thousandth of the energy we receive from the sun.

The history of solar radiation presents a great question. According to the straightforward theory of stellar evolution, early solar radiation should have been 30 percent less than it is today. If that had been the case, the earth's surface would have been frozen shortly after its formation and would have remained frozen, as explained above. One proposed explanation is that solar radiation was better utilized at early times. Out of nitrogen and hydrogen, the main components of the young earth's atmosphere, ammonia could have been formed. Ammonia serves as an excellent blanket. It makes it difficult for infrared radiation to escape the earth. This may have prevented an early, self-perpetuating glaciation.

It is known that the temperature of the earth was not always the same. Ice ages have occurred in the relatively recent past. It is less generally known that the earth has had extended ice epochs. These seem to last for a couple of million years. In the long periods of more than a hundred million years between ice epochs, there probably was no ice cover, not even on the poles. We are now in the midst of one of these exceptional ice epochs. Within an ice epoch, in turn, there are ice ages that last for a few tens of millenia during which glaciers cover much of what today are temperate zones. At present we enjoy the respite called an interglacial period, with relatively little ice cover left near the poles.

We live, therefore, in a twofold exception. It is exceptional that we happen to be in the midst of an ice epoch; and during this epoch it is exceptional that the polar ice caps should be as limited as they are now, less extensive than they were during the last ice age and less than they probably will be again in not so many thousands of years.

What has caused these changes? It may be that at various periods the earth utilizes solar radiation with various efficiencies. The reason may also be some slight variation in the earth's orbit, or some related astronomical cause. However, we cannot exclude the possibility that there are corresponding changes in solar radiation itself. We know of spectacular pyrotechnic displays on the sun which are connected with sunspots. They cannot be seen by looking into the sun with the naked eye, of course, because of the sun's brightness. There is a slight indication in the study of old plants or their remains that earth's temperatures and sunspot activity

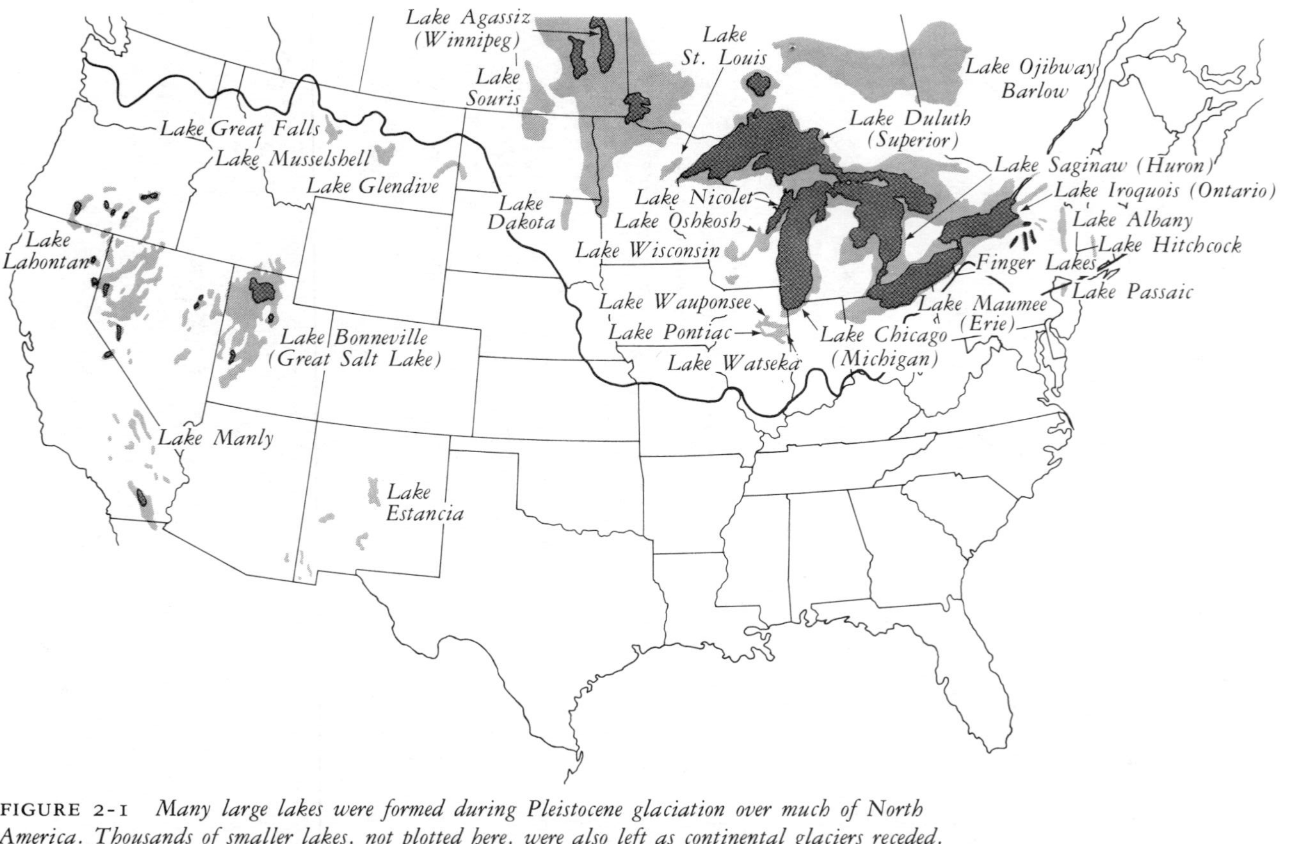

FIGURE 2-1 *Many large lakes were formed during Pleistocene glaciation over much of North America. Thousands of smaller lakes, not plotted here, were also left as continental glaciers receded. Major Pleistocene lakes are shown on the map in gray, their modern remnants in black. The region north of the continuous irregular line crossing the map was glaciated. Not shown are intermittent glaciers south of the area of solid glaciation, such as those in the American Rockies or the Sierra Nevada.*

may be related. Reasons for these changes are unknown, and the effect upon total radiation, if it exists, would seem surprising. It is obvious, however, that our understanding of the sun is incomplete.

The main facts of energy production within the sun are known, though, and on the whole they fit together like pieces of a puzzle. The conclusion remains that throughout the ages the variation of solar radiation has been moderate.

The stars and the sun derive their energy from thermonuclear reactions, which occur between nuclei at high temperatures. Atomic nuclei are positively charged and repel each other, and cannot get close to each other unless they move at high velocities. They reach these velocities only at high temperatures. Energy-producing reactions, therefore, occur only

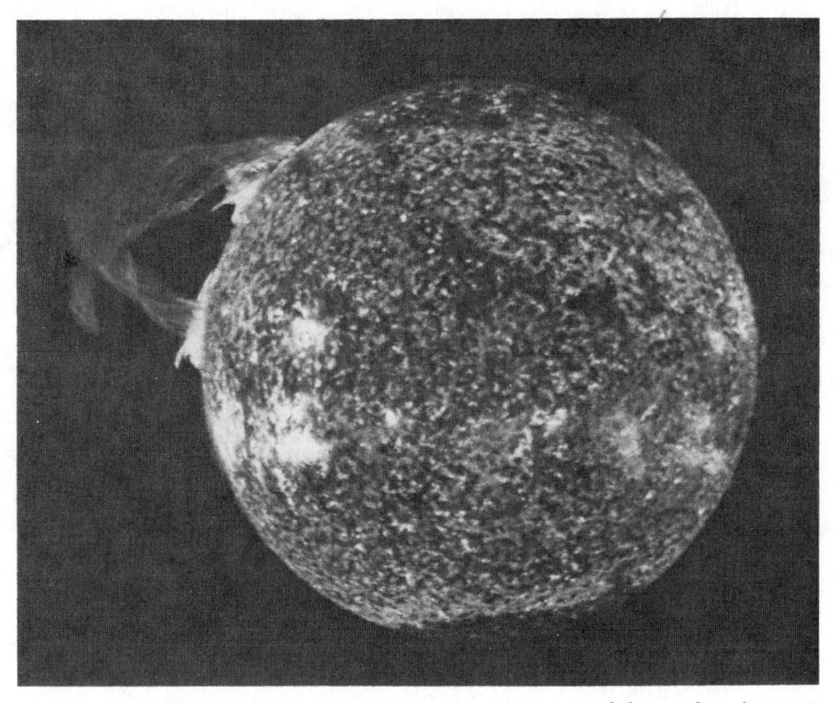

FIGURE 2-2 *Eruptive prominence of August 21, 1973, extended more than 400,000 kilometers above the surface of the sun. It was photographed with the Apollo Telescope Mount on* Skylab, *using appropriate light in the ultraviolet spectrum. On the scale of this picture the earth would be a dot somewhat larger than the period at the end of this sentence. (Naval Research Laboratory.)*

near the hot centers of stars, where temperatures are measured in millions of degrees. Near the center of the sun a twofold decrease in temperature causes the rate of energy production to drop by more than a factor of ten.

It is an interesting circumstance that there is a long delay from the time energy is produced until the time it reaches the surface of the star. In the case of our sun the delay is approximately one million years. Such a delay always occurs if the energy is transported by the method usually assumed: repeated emission and absorption of light in which the light performs what is called a "random walk," resembling the stagger of a drunken man whose every step is unrelated to the previous one. This mode of energy transport is the most obvious though not the only way. It is amusing to consider that when the first man built the first fire on earth, the solar energy that is reaching us now had already been produced and was on its way from the center of the sun toward its surface. This delay is one contributor to the steady nature of solar radiation.

The fact that solar radiation remains constant for millions of years shows that the energy production in the center of the sun is steady. Roughly speaking, the reason is that energy-producing reactions near the center of a star act like a thermostat. At first one might suspect that the thermonuclear reactions would be unstable, and would cause explosions. Indeed, the more energy is produced, the higher the temperature; and the higher the temperature, the greater the rate of energy production. Actually the effects of gravitation intervene, bringing stability. The more energy is produced, the more the star expands, and as a consequence it gets cooler. This process inhibits the excess production of energy. Conversely, if too little energy is produced, the star contracts, the temperature rises, and the deficiency in energy production is eliminated. Stability, therefore, is due to an apparent paradox: the more energy, the less heat.

The quantitative progress of thermonuclear reactions in the sun is quite complex, but the main lines can be stated in a simple way. By far the most abundant element in the sun is hydrogen. The protons that are hydrogen nuclei have the smallest charge and can approach each other more easily than other nuclei. Each proton suffers approximately one trillion collisions per second (10^{12} or a million times a million collisions). But even in ten billion years there is only a moderate chance that the proton would have undergone a nuclear reaction. Actually the only possible nuclear reaction is that as two protons approach, a positron (positive electron) and a neutrino are emitted, one of the protons is transformed into a neutron, and the proton and neutron are bound together into a deuteron or heavy hydrogen nucleus. We have discussed the formation of the deuteron shortly after the big bang. But at that time, protons encountered ready-made neutrons. This happens with high probability. In the sun, on the other hand, the formation of neutrons and their capture must all occur in one act; and formation of the neutron in the absence of the enormous streams of electrons and neutrinos available at the time of the big bang is exceedingly improbable. The more sluggish nature of the proton-proton reaction accounts for the moderate energy production of the sun and for its long life.

After the deuterons have been formed a complicated sequence of reactions is started, and in a relatively short time helium nuclei or alpha particles, containing two protons and two neutrons, are formed.

How do we know all this? In particular, how can we be certain about the extremely low probability of the initial reaction between two protons? The answer is, not by direct experiment but by reliable portions of nuclear theory. The history of the burning of deuterons has been checked by laboratory experiments. The first step, the formation of deuterons, is slow and so unlikely that a direct experiment may never be carried out.

Under these conditions it would be most desirable to find some further verification. We hope to accomplish this by observing neutrinos, chiefly those connected with the complex reactions in which the sequence of reactions beginning with the production of deuterons results in the formation of alpha particles. These latter neutrinos have more energy and are more easily detected. Neutrinos take only eight minutes to journey from the sun to the terrestrial laboratory. As yet we have no evidence for their presence. Experimenters have looked for them but have failed to find them, a result that led to some modifications in the detailed theory describing the thermonuclear reactions in the sun. There are now plans to find the neutrinos that accompany the formation of deuterons. Their number and their properties can be predicted within rather narrow limits. Meanwhile, the processes in even the star nearest and most important to us remain in doubt.

The story of the sun, incomplete as it is, has one consequence for the inhabitants of earth. The end of the world is *not* near. An early catastrophe involving the sun is most unlikely. A scientific description of the creation of our little human world should contain this important statement: We are still near the beginning of the act of creation. Probably our future here on earth will be a few times as long as the age of the oldest fossil. Gloomy stories we may hear about our dying planet are most likely wrong.

This statement is not necessarily optimistic. A potentially long future means, to me, a great responsibility. It is possible that we humans may eliminate ourselves from this long future, though this is not as likely as most stories, whether from poets, scientists, or futurologists, suggest.

Human imagination always has been attracted to the beginning, and even more to the end. It is probable that man and his peculiar descendants will be part of the future of our communal spaceship. Probably the portion of life now known as humanity will play the most important role in altering the face of the earth.

What kind of a face the earth will have, as millenia turn into millions and billions of years, is a very different question.

CHAPTER 3

BIOLOGICAL ORIGINS OF FUELS

In which life appears on earth, an event so surprising that it may well be called a miracle. Life's energy is derived from the sun. A very small portion of this energy is stored up as coal and petroleum, in a manner that experts debate but no one understands.

Almost all fuel used today has a biological origin. Coal, oil, and natural gas are derived from decaying life, mostly plant life. The living world, in turn, obtained its energy from the sun. The conversion of sunlight into a chemical form is, in itself, an interesting story from which lessons might be learned concerning the utilization of solar energy.[1] In the subsequent discussion we shall retain the old idea that all fossil fuels are indeed canned sunshine, the canning process to be explained by the phenomenon of life. There is, however, a different possibility which deserves, at least, to be mentioned.

In meteorites, which are fragments of old planets, we find hydrogen-carbon combinations that have some similarity to fossil fuels. Furthermore, some deep wells of natural gas have been found that seem to be better connected with vulcanism than with remnants of life usually found in sedimentary deposits. One theory holds that hydrogen and carbon combined to form sources of natural gas deep in the earth at a very early time in the evolution of our planet.[2] If this view is correct, natural gas might turn out to be in much more ample supply than we now suspect, provided we look for it in the right places. In this chapter, however, we shall pursue the older and more conventional ideas about fossil fuels.

As stated in the preceding chapter, our planet was formed approximately five billion years ago. Fossils that give evidence of well-shaped living organisms go back for only five hundred million years. Yet there is evidence of much older life on earth. During more than half of earth's existence there has been some form of life on it.

We have mentioned that evidence of such life is found in the form of peculiar chemical compounds that appear to be remnants of the building blocks of all living tissue, the proteins. It is assumed that early life existed in the oceans, and food for that life may have been something best described as manna from heaven. What was the origin of that manna?

The topmost layers of the atmosphere are exposed to ultraviolet radiation from the sun. This is only a small portion of solar radiation, but it is high-frequency radiation (invisible to the human eye as is low-frequency infrared radiation.) It arrives in big chunks, or *quanta*. Actually, all radiation consists of such quanta, but the higher the frequency, the bigger

---

[1] Melvin Calvin, "Photosynthesis as a Resource for Energy and Materials," *Science*, 184:375–381 (1974).

[2] Sandra Claflin-Chalton and Gordon J. MacDonald, *Sound and Light Phenomena, A Study of Historical and Modern Occurrences* (McLean, Va.: The Mitre Corporation, 1978).

these quanta, or energy bundles. In the ultraviolet spectrum, energy quanta become big enough to be absorbed by rather stable molecules and to excite them. Once the molecules are excited they are no longer stable. Most ultraviolet radiation gets absorbed in the atmosphere. The little that comes through is responsible for sunburn.

From this scant high-frequency radiation came manna, that is, molecules more easily capable of undergoing chemical changes. Too much radiation, and particularly too much ultraviolet radiation, would have destroyed these molecules again. These few high-energy quanta provided for the formation and at least partial preservation of more complicated molecular structures that, in turn, could serve as food for rudimentary forms of life.

Three billion years ago the atmosphere probably consisted of nitrogen, some hydrogen and ammonia, a very little water vapor, and carbon dioxide. The elements represented, carbon, hydrogen, oxygen, and nitrogen, are the chief constituents of living matter. All these molecules are stable (nitrogen, hydrogen, ammonia, water, and carbon dioxide, or N_2, H_2, NH_3, H_20, and CO_2); they cannot release energy by reacting with each other. Visible light does not affect them; that is why they are colorless. But they absorb ultraviolet light; in the ultraviolet one may say they are colored. Once light is absorbed by these molecules, chemical reactions can occur among them. In a few cases these chemical reactions result in more complicated molecules that have some energy content. This is the nutrient which I have called manna, which falls from the sky into the ocean.

No one knows how life originated on earth (or rather, in the sea). Was it an extremely improbable chance? Did the germ of life arrive from outside of earth? (This of course would not answer the basic question about the *origin* of life. It means only that we pass the buck.) In fact we can't even define life, much less state how it started. Following an ancient tradition one may say the origin of life was miraculous.[3]

But once it started in the oceans, it needed nutrients. The most plausible hypothesis is that those nutrients that rained down from the skies for a couple of billion years may have established a rich soup in which life could flourish and multiply.

---

[3] A great physicist defined a miracle as "an event with a probability of less than 10 percent." If that is correct, the existence of life is due to a miracle of a high order.

What I have said so far is hypothetical. What I will say now is quite fanciful. One may imagine that once monocellular living beings (or even smaller ones, because the cell is already a complicated little world) started to consume the nutrient they multiplied, and the first, potentially catastrophic population explosion may have occurred. An observer in the third billenium B.C. might have worried that competition in overpopulated oceans could result in the extinction of the life so recently arisen. But extinction was averted by a splendid discovery, and this discovery is not fancy but fact.

Of course, the discovery was not made by a conscious effort. It was, rather, made in the peculiar fashion in which most great and remarkable discoveries in the living world arise: by lots of trials and almost as many errors. The infinitesimal residue of successes is known as *evolution.* The name of the discovery was *chlorophyll.*

Chlorophyll is an extremely complex chemical substance designed to utilize sunlight much more efficiently than was the case in the production of manna. Indeed, the bulk of the sunlight penetrating the atmosphere to arrive at the surface of earth and ocean was utilized. The action of chlorophyll is understood only in general outline. No one has succeeded in duplicating that action in a test tube. So far, only chlorophyll within living plants is able to utilize sunlight in this peculiar and tricky manner. It is probable that the discovery of chlorophyll did not occur in one step. Chlorophyll may have been the final result of a lengthy evolution, but we can describe what it accomplished once it got into action.

Chlorophyll is green. It absorbs visible light in the red and blue parts of the spectrum, which means that it absorbs light quanta smaller than those found in the ultraviolet part of the spectrum. Chlorophyll itself, in its general chemical behavior, is almost as stable a chemical compound as nitrogen or water or carbon dioxide, but it has the remarkable ability to store most of the absorbed energy of a light quantum and wait for a second light quantum to arrive. It appears that this process is repeated altogether four times, though not necessarily in the same way. The energy that accumulates from the four light quanta is sufficient to compose, out of carbon dioxide and water, basic plant components and nutrients such as cellulose and sugar.

The importance of this discovery is clear. Chlorophyll can make use of most of the solar energy in the visible region. Furthermore, it can utilize that portion of the solar energy traversing the atmosphere of stable, colorless molecules that visible light can penetrate with little difficulty.

FIGURE 3-1 *An artist's conception of the land under the friendly skies of more than a hundred million years ago, when some solar energy was converted into fossil fuels. For the purpose of scale we should note that the pleasant animal shown in the drawing is not a swan but a brontosaurus. (George Bing.)*

Thus, chlorophyll made it possible for abundant vegetable life to exist on earth. But even more was accomplished. Not only was the body of the plant made out of carbon dioxide and water, but, in addition, oxygen was released in the course of photosynthetic plant growth.

Oxygen, in turn, made animal life possible, first in the sea and later on land. It has been said that animals are parasites on plants. This is true because food for animals comes from plants in a direct or indirect fashion. But it is also true because we owe to plant life the oxygen we breathe. It is probable that the earth's atmosphere developed a substantial oxygen content as recently as a billion years ago.

Having sketched the origin of the biosphere, the sphere of the living, we shall return to the production of fuels. It is of some interest to compare the total energy of these fuels with the total energy delivered from the sun to the surface of the earth during the last three billion years, the period in which some form of life existed on earth.

In describing these large amounts of energy it is convenient to use a *quad,* which is the equivalent of a quadrillion (a million times a million times one thousand) British Thermal Units (Btu). A quad equals the amount of energy consumed in the state of Iowa in 1973. Israel consumes approximately one-third of a quad a year. Our present world population utilizes about three hundred quads per year. To use another yardstick, a quad is released when 160 million to 170 million barrels of oil are burned.[4]

The total solar radiation reaching the outer boundaries of the earth's atmosphere is approximately four million quads per year. This is vastly more than the energy that is utilized. In a very crude approximation, one-thousandth of this energy, or four thousand quads a year, is stored by plants. This quantity varies in different periods. Furthermore, the energy of growing plants comes in part not directly from solar rays, but from decaying plants of earlier years.

This fraction of one-thousandth is composed of two factors. First, less than a tenth of the earth's surface is covered by growing plants. Second, the efficiency with which a plant converts solar energy into chemical energy is not much more than one percent. Even at this value, chlorophyll is performing well. The efficiency of producing manna was probably closer to one part in a million than the one part in a thousand that plants accomplish.

The total energy stored in known fossil fuels—in coal, oil, natural gas, tar sand, oil shale, and others—equals approximately one hundred thousand quads. If we include undiscovered resources, the figure could be as high as a million quads. The smaller of these two figures could have been produced by our present vegetation in as short a time as twenty-five years, the greater figure in two hundred fifty years. If you adopt as an intermediate figure four hundred thousand quads in fossil reserves, this would have been produced in a hundred years. This amount is less by a factor of twenty-five million than the energy available to plants in the last three billion years by the rays of the sun. How are we to explain this tremendous inefficiency?

A loss of two-thirds of the energy could be due to the inefficiency of converting vegetable matter into coal, oil, and the like. A factor of ten could be due to the circumstance that solar energy was less efficiently utilized in the past than in recent geologic ages. But in any case, we seem

---

[4]See, for example, *Monthly Energy Review* (U.S. Department of Energy, May 1978).

to be left with a residual factor of one million. This is a measure of the infinitesimally small probability that a living plant will eventually wind up as fossil fuel.

How do living plants turn into fossil fuels? I have gone to the best geologists and the best petroleum researchers, and I can give you the authoritative answer: no one knows. The circumstance that the transformation occurs in only a vanishingly small fraction of the vegetation makes the problem more difficult. It is by no means easy to track the chain of coincidences that eventually turns forests into coal seams or aquatic plants into oil deposits. It is not even certain that this transformation always proceeds in the same manner.

There is, for instance, one question of real practical interest that we must ask about fossil fuels. Coal is dirty and hard to transport. Oil is cleaner and much easier to ship. Yet, perversely, there is much more coal than oil. No one knows why.

We do know a few important differences between coal and oil. In plants, carbon and hydrogen combine in roughly the atomic ratio of one to two. In oil this ratio is virtually unchanged, but oil differs from the original living tissue in two ways. The atoms are linked together in a much simpler fashion, and most (but by no means all) of the atoms other than carbon and hydrogen have been eliminated. The product depends roughly upon the number of carbon atoms linked together. We obtain natural gas if the atoms are few, and in this case hydrogen atoms are more than twice as abundant as carbon atoms. Oil is the result if the number of carbon atoms is between five and a dozen, and tar or asphalt are produced in the case of even bigger molecules.

For coal the ratio of carbon to hydrogen has substantially changed. For lignite, the coal that used to be considered the poorest quality, the ratio of hydrogen to carbon atoms is somewhere between one and two. Brown coal has approximately an equal number of hydrogen and carbon atoms; whereas in the "best" coal, anthracite, the number of hydrogen atoms might be as low as one percent of the carbon atoms.

Both coal and oil, but *not* natural gas, include other atoms, such as sulfur, which is particularly troublesome since it contributes to air pollution. The amount of sulfur in coal may be as high as 6 percent or it may be well below 1 percent.

The generally accepted explanation of the difference between coal and petroleum (the collective name for gas, oil, and tar) relates to the role that oxygen plays in the transformations to which dead plants are subjected. In the case of coal oxygen has access to the decaying plant; in the case of oil it has virtually none. Oxygen combines more easily with hydrogen than with carbon, and therefore hydrogen will be more rapidly depleted if oxygen is present. This has led to the conclusion that coal has been formed from land plants, whereas vegetation in shallow seas yields oil.

Actually, "continental shelves" are found on the periphery of the continents. They are continuations, under the ocean, of the continents. Generally these shallow seas are rich in plant life. Oil found on dry land today was formed in a past geologic period when that land was part of a continental shelf.

Plants sinking to the bottom and being covered up by sediment are sufficiently isolated from oxygen by means of water and sand to preserve the rough ratio between carbon and hydrogen.

In general, sea plants are older and possibly more abundant than land plants. This is remarkable because coal seems to be a hundred times more abundant than oil. We must now consider some of the guesses as to how coal and oil may actually have been produced.

A usual assumption about the way coal was formed is that dead trees fell into swamps, that direct oxidation did not occur but some oxygen was around, and that bacterial action had a lot to do with the transformation from wood to coal. During this process some of the plant structure may be preserved, and it is sometimes still visible in coal. The popular notion is that the formation of coal usually occurs in tropical zones, but we have quite a bit of evidence that the process is going on in the sub-Arctic as well.

FIGURE 3-2 *Giant Fern of the genus* Pecopteris, *part of earth's vegetation three hundred million years ago in the carboniferous or coal-forming era, left the imprint of this frond in a thin layer of shale just above a coal seam in Illinois. (Smithsonian Institution)*

I have a favorite theory which I am sure is not correct in all cases, but may be correct in special instances. In the summer of 1958 I visited Alaska for the first time. It was near the summer solstice, and on my flight north of Anchorage into the wild, uninhabited regions of Alaska, I saw there was still a lot of snow on the ground. In many places the snow cover was pierced by rising plumes of smoke—a striking phenomenon. I asked what it was and was told that it was a forest fire. A forest fire in the vast, lonely stretches of Alaska is at first not noticed. As a rule it is never put out; it burns on and on. Winter comes and the fire is covered by snow. It may smolder all winter. Next spring the fire breaks out from under the snow.

You may notice that the situation I describe is similar to the one by which charcoal is conventionally produced. A big pile of wood is covered with earth and burned while it has limited access to oxygen. Perhaps some coal deposits are due to forest fires similar to the one I saw in Alaska.

We know only a little about the way vegetation became coal. For instance, we do not know whether we should consider a progression from lignite through soft coal to hard coal as slow evolution, or whether we should assume that these deposits result from different processes.

It seems that the formation of oil is an even more complicated process. The continental shelf where the original sea plants grew is shallow, its depth not over two or three hundred meters. Presumably, the plant remains were covered with sand, and during decomposition by anaerobic bacterial action the organic remains may have been linked to the sandy substrate, which could have consisted of remnants of seashells.[5] What is produced in this way is in fact not oil but oil shale.[6] It is an interesting substance, and we shall consider its practical uses in Chapter 7. Great quantities of oil shale are found in the Piceance Basin of northwest Colorado, and a sizable amount seems to be distributed in the form of a cup under the Dead Sea. In general, it is more abundant than oil.

It is believed that under pressure and at high temperatures oil is set free from oil shale. In order for oil to be obtained, the oil shale may have to be shifted to lower and hotter levels in the earth. The result of the heat

---

[5]This is not the usual fate of old seashells. As a rule they are compacted into limestone.

[6]"Oil shale" is a misnomer. It does not contain real oil and the rocky part of the substance is like limestone, a carbonate, while shale is a silicate.

is that the organic substance or hydrocarbon, called *kerogen* in this phase of evolution, is broken away from the sandy particles and becomes mobile oil. It can seep through porous, permeable layers of soil. It is lighter than water and is driven by the water to the top of those permeable layers. These collection points are called *stratigraphic traps,* and are of the greatest interest in oil exploration.

A typical trap is a rising permeable layer terminated by a fault, which is a surface along which all layers have been broken. The situation is illustrated in Figure 3-3. Regions labeled I, II, and III in the figure are nonpermeable rock. Between them, in a permeable layer, oil is trapped in the interstices of porous material above wet rock. Above the oil some gas may be trapped (not shown in the figure). The important point is that the original plant life may not have been close to the place where the oil is found. It may have seeped into the trap from a considerable distance, and, having collected, became ready for convenient recovery.

One particularly likely location for traps is a salt dome. Salt is deposited from evaporating sea water. Its horizontal layers are frequently covered up later by equally horizontal deposits of sand, which may subsequently be compacted into a more or less stony substance. The salt is less dense than the stone layers above it, and therefore the stone and salt would like to exchange places. This actually happens, with big masses of salt pushing, in the form of solid domes, into the layers that cover it. The process takes millions of years, and the rising salt bends and breaks

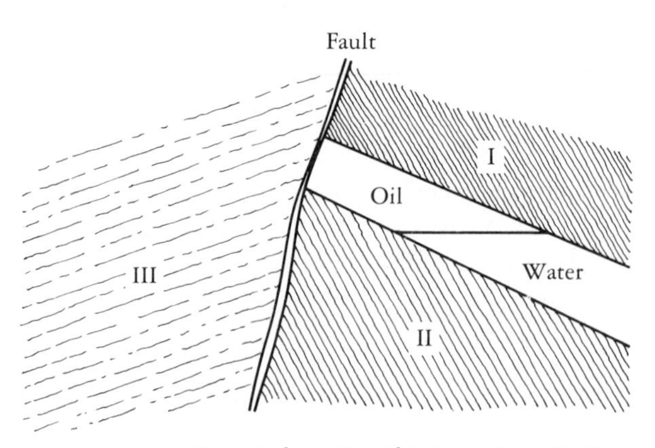

FIGURE 3-3 *A typical stratigraphic trap. Areas I, II, and III are nonpermeable rock.*

geological layers. A typical structure of traps is illustrated in Figure 3-4. The various shaded areas are different impermeable stony deposits. Oil and water may be found in the permeable layers. To make it more complex, one often finds natural gas on top of the oil layer or instead of it.

Salt domes and other stratigraphic traps can be located by geologists with relative ease. These traps influence and, in particular, reflect sound waves; therefore, they are sought rather in the way a doctor checks your lungs. But while the doctor taps, the geologist uses a high-explosive charge; and while the doctor listens with a stethoscope, the geologist uses a seismograph. Incidentally, although salt domes and other traps are easily located, the existence of petroleum near the salt dome is difficult to determine. Most salt domes are barren, and in general the expensive process of drilling is the only way to determine the presence of oil.

We have described a multistep process in the formation of oil. Much of what has been said is speculation. But it is certain that our description of the collection of oil in specific regions is correct. The involved nature of the whole process may account for the fact that so little of the ancient sea vegetation is found as usable oil.

One of the most important problems at present is the scarcity of oil—the temporary scarcity due to the Organization of Petroleum Exporting Countries (OPEC), and the long-term scarcity due to exhaustion of the limited amounts of oil actually available in nature. At this point I would like to discuss the latter.

There can be no doubt that considerably more oil will be found if we are willing to pay more for it. At present, oil exploration does not, on the

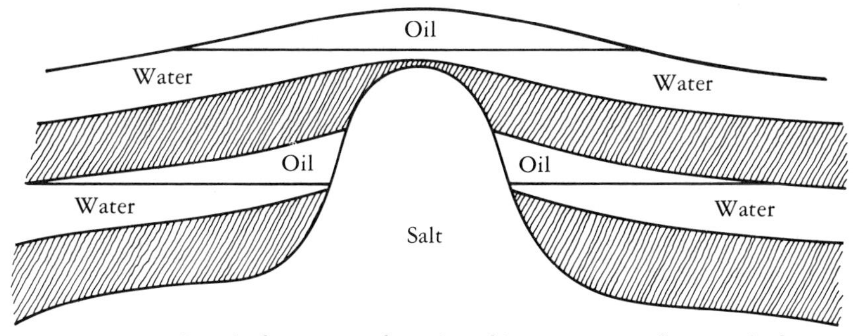

FIGURE 3-4 *A typical structure of stratigraphic traps surrounding a salt dome. The hatched areas are impermeable stony deposits.*

average, penetrate more than two kilometers below the surface. Oil has been found at greater depths, and it is likely that on a global scale we shall continue to find oil down to a depth of about six kilometers. Beyond that depth, pressures and temperatures rise to the point where conditions are changed. It is a matter of debate whether much more oil and gas will be found at greater depths. By going deeper than two kilometers we probably will find at least twice as much oil as is available in the layers so far explored.

I want to mention still another interesting possibility. Oil has been deposited and formed on continental shelves or in regions that were continental shelves in the geologic past. Beyond the continental shelf is the *continental slope,* the face of the "mountain" mass of a continent that rises sharply from the abyss of the ocean floor. Beyond the continental slope, however, you do not immediately find the abyss. Instead, you find the *continental rise,* that is, the detritus that has been carried down from the shelf. Figure 3-5 shows the arrangement. The continental rise, which is one or two kilometers deep, is the biggest storehouse in the world of sedimentary deposits.

We can actually see the process of formation of the continental rise. Earthquakes shake down material from the continental shelf and this material slides down the steep continental slope onto the continental rise.

There are salt domes below the continental rise, and near these salt domes oil has been found on at least one occasion. One possible answer to

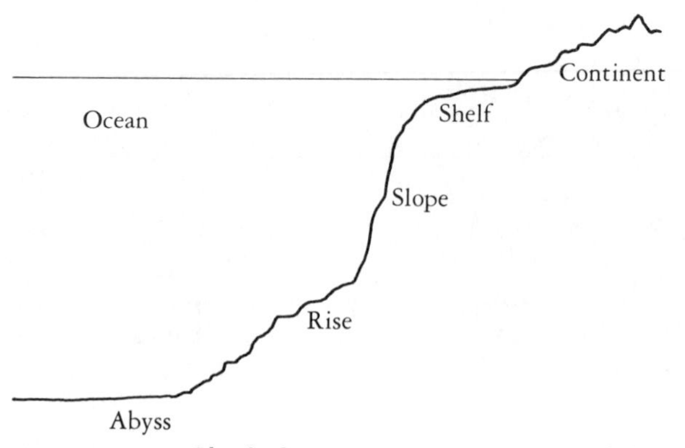

FIGURE 3-5 *Sketch showing approximate contours of the ocean floor where sea and land meet. In reality the slopes are not as steep as the figure indicates.*

the question of why we have found much more coal than oil is that there may be as much oil as coal in places where we have not looked as yet. One must realize, however, that if great quantities of oil actually should be found below the continental rise, then new, expensive methods using remote control will have to be used to recover it. This conclusion is a somewhat meager, somewhat distant, practical result of a long discussion.

It is obvious that at any probable rate of consumption, the earth's fossil fuel supplies will be effectively exhausted within the forseeable future. In the case of oil this may well happen in a few decades, though it could take a century, or if we are very optimistic, even a little longer. Coal will last for a few centuries, but even with coal the end is in sight.

This does not mean that, in the long run, we face a necessary cutback in energy consumption. Nuclear fuels will be ample for many millenia and may serve for a period comparable to geologic ages. Solar energy is both ample and, in a practical sense, inexhaustible. In time, the use of these energy sources will certainly become a practical reality.

But fossil fuels filled a most important historical role. They could be used easily and cheaply, and thereby they actually made it possible for the Industrial Revolution to proceed rapidly. As a result of the Industrial Revolution, new ways were found to satisfy our needs. There is no reason to assume any narrow limits for future energy resources, which are clean, ample, and lasting.

PART II

USE FROM THE RECENT PAST TO THE NEAR FUTURE

CHAPTER 4

CONVENTIONAL USES FROM PROMETHEUS TO THE OIL EMBARGO

In which a clever biped utilizes fire and turns into man. Energy is used to survive the ice ages in some degree of comfort; more energy is used to establish civilized communities. Finally, proliferation of energy in part of the globe is connected with the Industrial Revolution, coal, pollution, electricity, oil, and the oil shortage. The stage is set for asking many questions and receiving, perhaps, a few answers.

It is said by way of definition that man is an animal that uses tools. One might say with greater justification that man is an animal that uses energy. Many animals make use of objects: ants and bees build their cities; apes use stones and nuts and sticks in ways that are vaguely similar to the use of objects by stone age man. But only man uses fire.

The first use of fire occurred half a million years ago. The cultural shock must have been greater than that connected with the dawn of the atomic age. That ancient shock is still remembered in legends such as that of Prometheus, who stole fire from the gods and brought it down from heaven. I like to think that this particular legend is significant: man's utilization of fire was considered criminal, and Prometheus was cruelly punished. In this special story, at least, energy is connected with guilt. Are we reliving this ancient experience early in the nuclear age?

We know quite accurately how much energy man consumed before the use of fire. It was equal to the caloric value of the food that sustained him. Before 500,000 B.C., the daily per capita consumption was ten thousand British thermal units.[1] Most people would guess that the per capita energy consumption today is vast compared to that of our pre-Promethean ancestors. Actually the modern American consumes about one hundred times as much as our prehuman forebears, that is, a million British thermal units per day.

The use of fire may be identified, in an arbitrary way, with the origin of man. It seems that man's brain volume doubled at about the same time as fire began to be used. Mankind in that time certainly was a "wild" species, probably numbering well under a million.[2] At that period man did not put more pressure on the environment than other animals did.

---

[1]A British thermal unit (Btu) is the energy needed to heat one pound of water by one degree Fahrenheit. For physicists it is useful to remember that one Btu is very little more than 10^{10} times an *erg,* the unit in which physicists measure energy, which is twice the kinetic energy of one gram moving with the slow speed of one centimeter per second.

[2]I like to estimate the human population with the help of the formula

$$N = \frac{200 \text{ billion}}{2030-T}$$

where N equals population and T equals date (A.D.) In years B.C. the value of T is negative. For 500,000 B.C. this calculation gives a population of 400,000. The formula has no more theoretical justification than the popular exponential formula for population increase, according to which population should increase each year by a given percentage. However, it is in better agreement with recorded population statistics. It predicts a real population explosion for the year 2030, which means that the formula must break down before that time. Fortunately, a slowdown is already apparent.

FIGURE 4-1 *Sketch of a pseudo Greek vase. (George Bing.)*

From the most rigid point of view of an uncompromising environmentalist, that was the state of paradise. Not even the extremists in the environmental movement are prepared to return to that Garden of Eden.

With the coming of Prometheus paradise was lost, although not rapidly and not in a drastic fashion. One may crudely estimate that with the discovery of controlled fire, per capita consumption of energy doubled. It was still only 2 percent of present consumption by an average United States citizen. The additional fuel went for heating, which was exceedingly useful, particularly in the ice ages. Those drastic changes in climate must have been a little less terrible to man than to his fellow mammals. Cooking was of course another use for fire. Becoming human and playing with fire meant that our ancestors suffered less from cold and from hunger. By 10,000 B.C., their numbers probably had increased into the millions.

A little more than ten thousand years ago, with the ending of the latest (but probably not the last) ice age, a new revolution occurred which transformed man into a social animal. This revolution was man's extensive exploitation of the living world, both animal and vegetable, around him. This development probably was accompanied by a further increase in energy consumption. By that time an individual living in an "advanced" community may have consumed 5 percent of the present American average. This rate of energy consumption may not have changed substantially until the approach of the Industrial Revolution. It still holds for 50 percent of the people who live in the underdeveloped part of the world, called hopefully the developing world. Indeed, before 1950, in equatorial Africa energy consumption remained on a pre-Promethean level.

The general use of domesticated animals and the introduction of systematic agriculture was completed in the time we call the Neolithic age. Extensive use of herds and intensive agriculture began to leave their mark on the environment. Overgrazing occurred thousands of years before Christ. Agriculture, however, improved the land in many different areas: along the Nile, in Mesopotamia, along the Indus and the Wang Ho Rivers.[3] Irrigation turned arid land into fertile fields. The change in environment was not always for the worse.

It is remarkable that in the ensuing ten thousand years energy consumption and the standard of living changed so little for the average man. Instead, another tremendous change took place, which may best be called civilization.

Neolithic man had to have a social organization, at least on the tribal level. Society later assumed the form of a pyramid. The obvious positive aspects of this change were the development of writing (and the beginning of recorded history) and a possibility for a rich life and perhaps a good one near the top of the social pyramid. The negative aspects were more obvious at the bottom of the pyramid. The bottom of the bottom was some form of slavery. Kings, slaves, and even philosophers agreed that slavery was the necessary foundation of human society.

The most important aspect of Christianity may have been that it was a peaceable modification of the slavery system, a gentle nonviolent revolution that persisted for many centuries in its drive to make men more nearly equal. This drive succeeded in part. Even today success is obviously incomplete. It is doubtful whether it ever can be completed. Yet the revolution persists.

These considerations may seem like philosophy rather than history. What place do they have in the history of energy?

Professor Lynn White in his book *Machina ex Deo* gave the answer. The Christian revolution against slavery gave rise to the idea that man's labor is honorable. In the millenia in which slavery was taken for granted, ideas were considered ennobling. But bodily work, other than military exploits, was regarded as demeaning. It remained to the western Christian monastic orders, for instance the Benedictines in the sixth century A.D.,

[3]Silting and uncontrolled floods earned the Wang Ho its age-old title, "China's Sorrow."

to introduce respect for the work of human hands. *Ora et Labora,* putting work next to prayer, was the slogan that changed the world.

If work was to be respected, it was natural that it should be enhanced with the help of machines. In this curious yet practical fashion did the new form of belief in God lead, through the idea of brotherhood of men, to the use of machines and, in the course of many centuries, to the Industrial Revolution.

Machines came to western civilization not because Europeans were more clever than others, but because they had a greater desire for machines. Machines needed energy. So in western Europe, at the dawn of the Industrial Revolution, more energy was used than in other places. Human muscles were then better supplemented by machines than at any earlier time. By the middle of the eighteenth century, energy consumption in western Europe may have reached 10 percent of the modern American standard.

That the reduction of human labor was due primarily to desire rather than to inventiveness is emphasized by the fact that in the Middle Ages none of the labor-saving devices originated in Europe. The horse had been used in warfare for more than a thousand years. The harness originally used in the West, however, made it impossible for a horse to draw a heavy burden; the harness would have choked the horse. A better harness, which utilized the horse as a working animal, came to Europe from China.

A second example is the waterwheel. It was known to the Greco-Roman civilization, but its use became widespread in medieval Europe.

A final example is the windmill, which came originally from the windswept plateau of Persia. By the fifteenth century windmills were common in Europe and Don Quixote could fight these monsters, which were actually harbingers of the great change that was to come to Europe.[4]

The true start of the Industrial Revolution, conventionally placed at the end of the eighteenth century, had a number of other well-known antecedents. One was the exploration of the globe by ships from Europe. Another was the Copernican revolution and the birth of modern science. A third, less recognized, was closely connected with energy.

In the eighteenth century England ran out of firewood. It became necessary to use a substitute that was abundant and dirty—coal. Persons concerned about the environment were against it. In some places the burn-

[4]Amory Lovins, a modern Don Quixote, is a passionate advocate of wind power.

ing of coal was banned; users were even threatened with the death penalty. But coal was needed, and it was used.

Coal burns with a hotter flame than wood. It could be used in producing iron from iron ore. One may claim that the eighteenth century was the real beginning of the iron age. Prior to that time the cheapest metal was bronze. That was the reason cannons had been made of bronze rather than of iron. What historians call the beginning of the iron age corresponds to the introduction of iron in small quantities as weapons, especially swords. Iron was in fact the secret weapon of the conquering Hittites. The art of the smiths (an honorific designation) was to use charcoal and skill to produce the early masterpieces of metallurgy. The use of iron in beautiful medieval grillwork also indicates that iron was a valuable material on which care and art were lavished.

This attitude toward iron changed in the eighteenth century. Iron, a strong and abundant metal, became cheap. This was one of the conditions which made the Industrial Revolution possible.

According to textbooks, James Watt's steam engine is the landmark symbolizing the beginning of the Industrial Revolution. This notion is not mistaken. In 1765, when Watt made his first invention, iron was readily available. Watt's engine was used almost exclusively to pump water out of coal mines, the coal being used to a great extent to produce iron. Thus the steam engine was both the consequence and the accelerating force of the Industrial Revolution.

Coal became the main fuel of the industrial society in the nineteenth century. Energy consumption rose in the advanced countries to almost one-half of present American figures. At the same time, a price was paid in pollution. London's pea-soup fogs were caused by indiscriminate burning of coal. Only in the middle of the twentieth century, after the relationship of a higher death rate with the "killer" London fog had been observed, was the burning of coal effectively regulated. Those famous fogs of London are now a memory of a past but not completely golden age.

In America the history of energy consumption took a course quite different from that in the rest of the world. At present the average American consumes a little more than twice as much energy as the average European. (Incidentally, the per capita consumption is remarkably similar in eastern Europe and western Europe.) This extravagance is not a recent phenomenon, nor is it an exclusive result of industrialization. It is a fact deeply rooted in American history.

According to *Energy Crisis in Perspective* by John C. Fisher, based on figures of the United States Bureau of the Census, the per capita consumption of energy in the United States was remarkably high as early as 1800. At that time the Industrial Revolution certainly had not reached the shores of the New World. Even in its cradle, England, industrialization had barely started. Yet energy consumption in the United States at that time was as much as 40 percent of the modern figure per person in the United States. In relation to the energy consumption of the rest of the world, America took much more than its share, substantially more than it does today.

If we look in greater detail, we find that most American energy consumption was due to the burning of firewood. The portion of the United States populated by the white man was hemmed in by forests. It was considered an advantage to "clear" the forest and burn the wood.

Another big contributor to energy consumption was animal fodder, which was generally turned into horsepower. Americans took possession of a big, new, rich territory. One is tempted to say that our ancestors squandered this richness, that we are the spendthrift followers of an even more spendthrift early tradition. But such a statement may be unjust.

One may argue that it was the abundance of the country that allowed the flood of immigrants to attain a better life. It was this same abundance which permitted the exceptional growth of the Industrial Revolution in this fertile environment. It is not clear that America was depleted by spending more. In a peculiar human sense the country was replenished.

This view of our forebear's energy consumption raises a question about the environment to which we must return in subsequent chapters. But let us recognize here that abundant and perhaps excessive use of energy is only partially characteristic of industrialization. It also happens to be an important part of American history.

Abundant use of energy is not the only way in which the Industrial Revolution took a characteristic course in America. Emphasis on transportation and communication is related to the great size of the country. Apart from other important differences, such as early development of mass production, additional qualitative effects are connected with energy.

It was only in the 1880s that the use of coal, which was in abundant supply, overtook the use of firewood in the United States. But by this time two additional changes had been initiated which were to have a great influence on the future of energy. One was Edison's work on electric power; the other was the use of petroleum. In both of these fields the United States led the way.

Apart from its versatility and cleanliness, electric power is particularly important because it provides a cheap method for transmitting useful energy over considerable distances. Electric power is, of course, in no sense a source of energy, but only a way of utilizing energy.

Among early sources of electric energy, water power was particularly important. Steam plants, fired by coal, became widespread in the first half of the twentieth century. Then followed the substitution of oil and gas for significant amounts of coal. At present a most important question is whether nuclear energy, geothermal sources, or some form of solar energy can replace the fossil fuels in generating electricity. These questions have great bearing on the causes and possible cures of the energy crisis.

But some historic trends should be mentioned here. In the twentieth century there was a tremendous increase in the use of electric energy in the United States, due to three different causes. First, total energy consumption increased approximately sixfold. Second, the fraction of energy used in generating electricity increased approximately fourfold. But in addition, the efficiency of converting fuels into electric power had a similar increase, from less than 10 percent to as much as 40 percent in modern plants. Altogether, in the first three-quarters of the twentieth century, electric output increased almost a hundredfold. In the same period the population increased only threefold. From a minor contributor, electricity had developed into a major part of our human environment.

In considering electricity, emphasis is often placed on the circumstance that in average American plants (not all of which are modern) only one-third of the energy becomes electricity whereas two-thirds appears as waste heat. The waste is not as unusual as it would appear. Every use of energy is accompanied by some waste—sometimes less, but in the majority of cases more than with electricity.

The development of electricity in the United States and other industrialized countries has been similar except for one important point. In the last twenty-five years electrical interconnections have proliferated to a greater extent in the United States than anywhere else. Interconnections mean that electricity produced in one place can be utilized at a distance of hundreds of miles. In particular, the needs of one consumer can be satisfied from a multiplicity of sources. This would seem superfluous but in fact it is exceedingly useful. The bigger the system, the greater the efficiency of utilization.

Local blackouts can be avoided by obtaining help from a more distant generating station. In addition, fluctuation in demand can be evened out, at least to some extent. The cost of generating electricity is primarily the cost of fuel and capital investment. Capital investment has amounted to as much as one-third of the cost during the last few decades, and will become even more important in the future. Permitting valuable equipment to stand idle is uneconomical. Even though some energy is lost in transmitting electricity, it still pays to use the cooperation between generating stations for the purpose of evening out fluctuations in load. Interconnections began proliferating in 1950. The result has been a considerable lowering of the electricity bill for the individual user.

But interconnections have disadvantages as well as advantages. They prevent small blackouts but, as was demonstrated on the night of November 9, 1965, they can produce a blackout on a tremendous scale. What happened then is of interest, because in a time of electricity shortage the circumstances could easily be repeated unless appropriate precautions are taken.

The interconnected system CANUS (which stands for Canada-United States) linked New York, New England, and parts of Canada. Power was generated in New York, Boston, Niagara Falls near Buffalo, and other stations including a few in Canada. A Canadian station had to be shut down for maintenance. There was plenty of electricity in the United States on which Canadians could draw. Unfortunately, the circuit breakers on the CANUS interconnection were set at an unnecessarily cautious and low level. Circuit breakers are necessary to prevent destruction of the line by excessive current. In this case, by mistake, the permitted current was about one-half what it could have been. The Canadians forgot this when they drew on the American supply at approximately 5:15 P.M., when electricity consumption was near its peak.

The circuit breakers went into action and the interconnection was broken in less than a second. In Toronto, lights went out and subways stopped. But Toronto was also hooked into Chicago, so service was restored in half an hour. In the United States the consequences were much worse.

The region near Buffalo suddenly had an excess of electricity and a deficiency of *load,* that is, there was suddenly too little demand for electricity. An electric generator receives its energy from a mechanical source that may be a turbine driven by water or steam. The revolutions of the generator are, however, slowed down by the process of generating electricity, that is, by the electric load. If the load is decreased the generator speeds up. Too much increase in speed damages the generator; therefore these generators are, in turn, protected by equipment that disconnects them automatically if the speed increases excessively. Furthermore, a generator that is once shut down cannot be started up again for a period perhaps as long as half an hour. Near Buffalo the decrease of electric load led to the temporary elimination of a much greater chunk of generating equipment. Within a few seconds the excess generating capacity turned into a big deficit.

Now the action was up to the dispatcher of the biggest generating unit in the area, Consolidated Edison of New York.

There are provisions for incidents of the kind that developed. Electrical generating capacity is kept available in the form of *spinning reserve.* These are generators that are rotating or spinning at a low power output from which power can be delivered when the load increases on very short notice. The dispatcher saw that there was plenty of spinning reserve; therefore he did not choose to reduce the load by accepting a temporary blackout or even a brownout in New York, or in other regions in New England that were beginning to draw heavily on Con Ed.

Not reducing the load was a mistake. The big Ravenswood plant in New York had a faulty valve. When more spinning reserve was called in power was temporarily restored, but because of poor operation at Ravenswood energy stored in the form of hot steam was exhausted within a few minutes. Due to this mistake and a few similar ones, the whole northeast electrical system collapsed. Tens of millions of people went without electricity for up to eighteen hours. The consequences could have been serious except for good weather, bright moonlight, and the good spirit of the New Yorkers.

Air traffic had to be diverted from airports. A few accidents did occur but, remarkably enough, the crime rate dropped. Apart from spoiled food in nonfunctioning refrigerators and other similar losses, the one marked effect was a sharp temporary increase in the birth rate nine months later.[5]

In December 1965, Governor Rockefeller of New York asked me to study the causes and cures of the blackout. It was a most interesting exercise. The original cause on the international border was not found for five days. Thirteen minutes had elapsed between that cause and the widespread blackout. For at least the first six or seven minutes effective preven-

[5]Twelve years later, in July 1977, a blackout in New York was accompanied by extensive looting, the jailing of 3,500 people, and damage exceeding a billion dollars.

tative steps could have been taken. The cause of the blackout was not weakness of the electrical system, but rather the unavailability of sufficient information.

It is less generally known that another blackout occurred on June 6, 1967, the day the Six-Day War started in the Middle East. At that time the New Jersey–Pennsylvania–Maryland system collapsed. It happened in the daytime, and while industry was heavily affected, individuals were inconvenienced to a lesser extent. The Six-Day War crowded this occurrence out of the news, but its course was very like that of the CANUS blackout. The original cause was a human error that led to temporary overproduction of electricity. Again, the final blackout could have been avoided during a period of many minutes had sufficient information been available.

The recommendation that some of us made after the 1965 blackout was that dispatchers be helped with elaborate computing equipment. Well-constructed computers do not forget the setting of circuit breakers or the state of faulty valves. Furthermore, computers can be asked to calculate within a second what the effect would be of various courses that a dispatcher may choose: What would happen in case of dropping load causing a blackout or a partial blackout? What would happen if spinning reserve were called in from this or another source? In the months following the CANUS blackout, electric utilities were unwilling to deploy computers for purposes of control. It was claimed that the computer can never be more clever than the human (a debatable claim). It was not recognized that in an emergency computers work faster and more reliably than humans. Decisions should indeed be left to the dispatcher; but computers can display the situation so that faults are detected in five seconds rather than five days.

Now that our electricity supply will have to operate on a narrow margin, such computer arrangements are becoming much more urgent. A beginning has been made toward introducing them.

Utilities are squeezed in three different ways at present. One is regulations, a second is rising fuel costs, and a third is rising capital costs. In the last few years, one partial solution has helped with the third item—the widespread introduction of gas turbines.

Gas turbines use expensive fuel and use it with the moderate efficiency of 30 percent. They are cheap and rapidly installed. They utilize technology developed in connection with jet planes.

In a steam-driven generator the working fluid, steam, is separate from the burning fuel, and heat-transfer surfaces have to be introduced. Combustion chambers and other pieces of apparatus are relatively massive and expensive. In a gas turbine, air is compressed by the action of a propeller. Fuel is injected into the compressed air and burned. Temperature and pressure are raised and this gas now drives a second propeller. Part of the energy is transferred back to the first propeller, which compresses the air. The rest of the energy is used to generate electricity. This is precisely the same principle and almost the same apparatus as the engine used today in big passenger planes. The difference is that in airplanes the excess energy goes into an airstream, or jet, which is ejected from the back of the engine, helping to push the plane ahead.

Gas turbines are ideal for providing the "peaking power" in electrical plants, that is, making electricity available at low capital cost for one or two hours a day. If they are used for longer periods their fuel consumption becomes much too expensive.

Another development is the recognition, after undue delay, that in at least some cases adverse environmental effects must be eliminated. In this connection electricity plays a special role. From the user's point of view electricity is clean, but at the point of production it may be very dirty indeed. This situation makes it possible to establish and to enforce proper environmental standards. It is much more difficult to look into the chimney of every household than into the few smokestacks of the electric

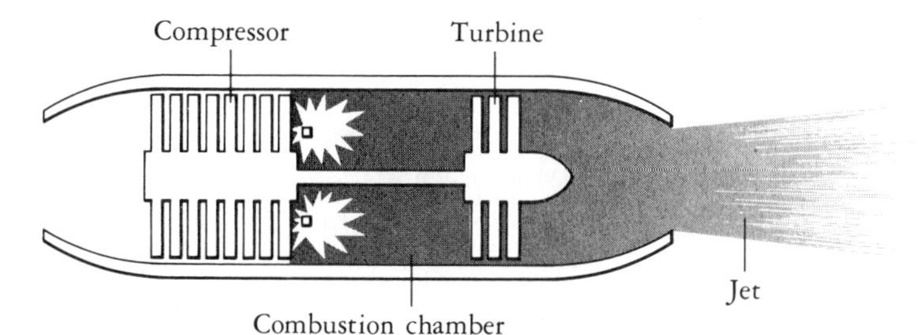

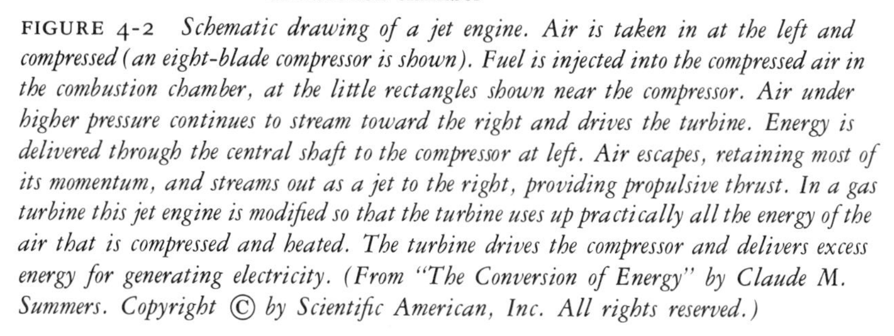
FIGURE 4-2 *Schematic drawing of a jet engine. Air is taken in at the left and compressed (an eight-blade compressor is shown). Fuel is injected into the compressed air in the combustion chamber, at the little rectangles shown near the compressor. Air under higher pressure continues to stream toward the right and drives the turbine. Energy is delivered through the central shaft to the compressor at left. Air escapes, retaining most of its momentum, and streams out as a jet to the right, providing propulsive thrust. In a gas turbine this jet engine is modified so that the turbine uses up practically all the energy of the air that is compressed and heated. The turbine drives the compressor and delivers excess energy for generating electricity. (From "The Conversion of Energy" by Claude M. Summers. Copyright © by Scientific American, Inc. All rights reserved.)*

plants. Environmentalists may take notice that small is not necessarily beautiful or clean.

If there is a development of greater recent historical importance than electricity, it is petroleum. It started a few millenia ago in the Persian Gulf where combustible material was lying around. Petroleum came to have great historic importance in the seventh century A.D., when burnable tar mixed with lime and sulfur was used as Greek fire to burn the Arab fleet that invaded Constantinople. But it was not until the second half of the nineteenth century that petroleum from Pennsylvania entered the market as a regular fuel. At that time the depth of an oil well was twenty meters. Today the average depth is two kilometers. Even so, the development of technology and savings connected with big-scale operations have resulted in a steady decline of the cost of petroleum and its products. This statement may sound peculiar when made in the midst of an energy crisis that is, in fact, a petroleum crisis.

The reasons that oil and later gas became exceedingly popular, first in the United States and later throughout the world, are threefold. Oil is easily produced, it is cheaply transported, and the equipment that turns oil into useful energy can be readily available. In all these phases of petroleum production, transportation, and utilization, the world has invested several hundred billions of dollars.

Now the word has spread that we are running out of oil. This is only partly true. Known reserves will suffice for fifteen years. Finding oil for the more distant future has little justification on purely economic grounds, particularly when the interest rate is high and capital is hard to obtain. On the other hand, exploration for new oil will result in production no less than five years hence. What is practical is squeezed between two uncomfortably narrow margins. While it is unprofitable to plan more than fifteen years ahead, it is impossible to obtain supplies from new sources on land in less than five years and from sources on the continental shelf in less than ten to fifteen years.

This situation was further aggravated by environmentalist demands that caused dirty coal to be replaced by not quite so dirty oil. The oil spill in the Santa Barbara Channel has limited the hopeful oil exploration on the continental shelf of America. Under these circumstances the power of decision was shifted into the hands of Arabs who control more than half of the world's known and developed oil reserves.

In the Gulf of Persia the cost of oil production, including ample allowance for writing off investment was 15¢ a barrel. In the United States in 1970 it was almost twenty times as much. Up to the end of the 1960s the United States restricted its oil production to less than full capacity. At the same time, few new oil fields were developed. This was not due to a lack of prospective fields. It was due to the availability of much less expensive oil in the Middle East, Africa, and some other places. It was also due to a strong belief among American economists in the principles of free trade.[6] By 1970 or 1971, United States production had reached capacity, and further expansion was greatly impeded by reduced activity in exploration. As long as the United States had some reserve capacity, the price of oil was regulated by the marginal price of American production, which was $2 to $3 per barrel. After 1971, this limiting price was no longer effective and the Organization of Petroleum Exporting Countries (OPEC) was free to make the decisions.

There were two further reasons why no rapid increase in the price of oil was expected. One was the rule of economics that price-regulating cartels cannot be established by producers of raw material. This law, as so many other laws of economics, proved to be a fable. Within a short time Arab economists (who are not, as their American colleagues are, dedicated to free trade) caught on to the idea that cartel politics could pay enormous dividends.

The second reason was stronger. Arabs reputedly cannot agree with anybody, least of all with other Arabs. But this limitation faded in the Yom Kippur War. The war afforded the Arabs a splendid opportunity to get together under Saudi leadership and establish an oil price of ten dollars or more per barrel. As a result, a hundred billion dollars flowed annually into the treasuries of OPEC, an organization established in the 1960s to prevent further *decrease* in the price of oil. Of this hundred billion dollars, half was spent in industrial countries. The other half represented an unprecedented rate of accumulating financial power in the form of liquid capital.

This lopsided financial situation could not and did not last. The income continued. Expenditures of the oil-exporting countries rose but, particularly in the countries on the western shore of the Gulf of Persia, did not catch up with earnings from oil. The industrial countries were hit by

---

[6] I am tempted to raise the question whether adherents of free trade are driven by enlightened arguments or by religious fervor. Free trade and politics are inseparable, and at present the question is unanswerable. At some future date the answer may seem obvious.

economic conditions that were politely called a recession. Not all the oil that could be produced for 15¢ a barrel could, in fact, be sold for $12. Even so, the demonstrated advantages of OPEC were strong enough to hold the cartel together, although it became a little less monolitic with the passage of time.

This account brings us up to the present situation. The worldwide energy shortage is least severe in the United States, which imports approximately one quarter of its energy. The situation is worse in western Europe, where coal mines have been shut down and where approximately 60 percent of the fuels come from abroad. Japan's spectacular success in industrial operations was brought to a halt by its almost total dependence on foreign energy sources. The developing world, which relies for its scarce energy supply on imports, which must keep pumping water for irrigation, and which needs oil products as fertilizer, is probably in the greatest difficulty. The increase in oil consumption there came to a halt and this meant postponement of hope for rapid development. Just maintaining the flow of essential oil caused severe economic problems.

One truly populous and extremely poor country that is an exception in this gloomy picture is Indonesia. It produces clean, valuable oil and has therefore gained the opportunity for development to help improve the lot of more than a hundred million people. A similar situation may develop in Nigeria.

The story of energy must include a special word concerning natural gas. This energy source differs from oil in two practical respects: it is invariably clean, and it is not easily transported across the ocean. In the first half of this century, natural gas was underutilized in the United States. The appropriate regulatory agency, the Federal Power Commission, has helped in encouraging construction of pipelines, and in this manner natural gas has become a major source of energy in America. More than one-fourth of our energy comes from natural gas.

The influence of the Federal Power Commission has become quite different in the last few years. Gas appearing in interstate commerce had been regulated at a price corresponding to $1 per barrel of oil. This ceiling was raised to $3 per barrel and later, for new gas, to $8.50. Nonetheless, gas remains less expensive than oil, which means that it is eagerly used, it is apt to disappear from the market, and it is becoming rapidly unavailable. The consequences are particularly dangerous for such a region as southern California. Weather conditions in that part of the country are such that combustion products will hang around, and Los Angeles has to stew in its own juice. Clean fuel, and particularly natural

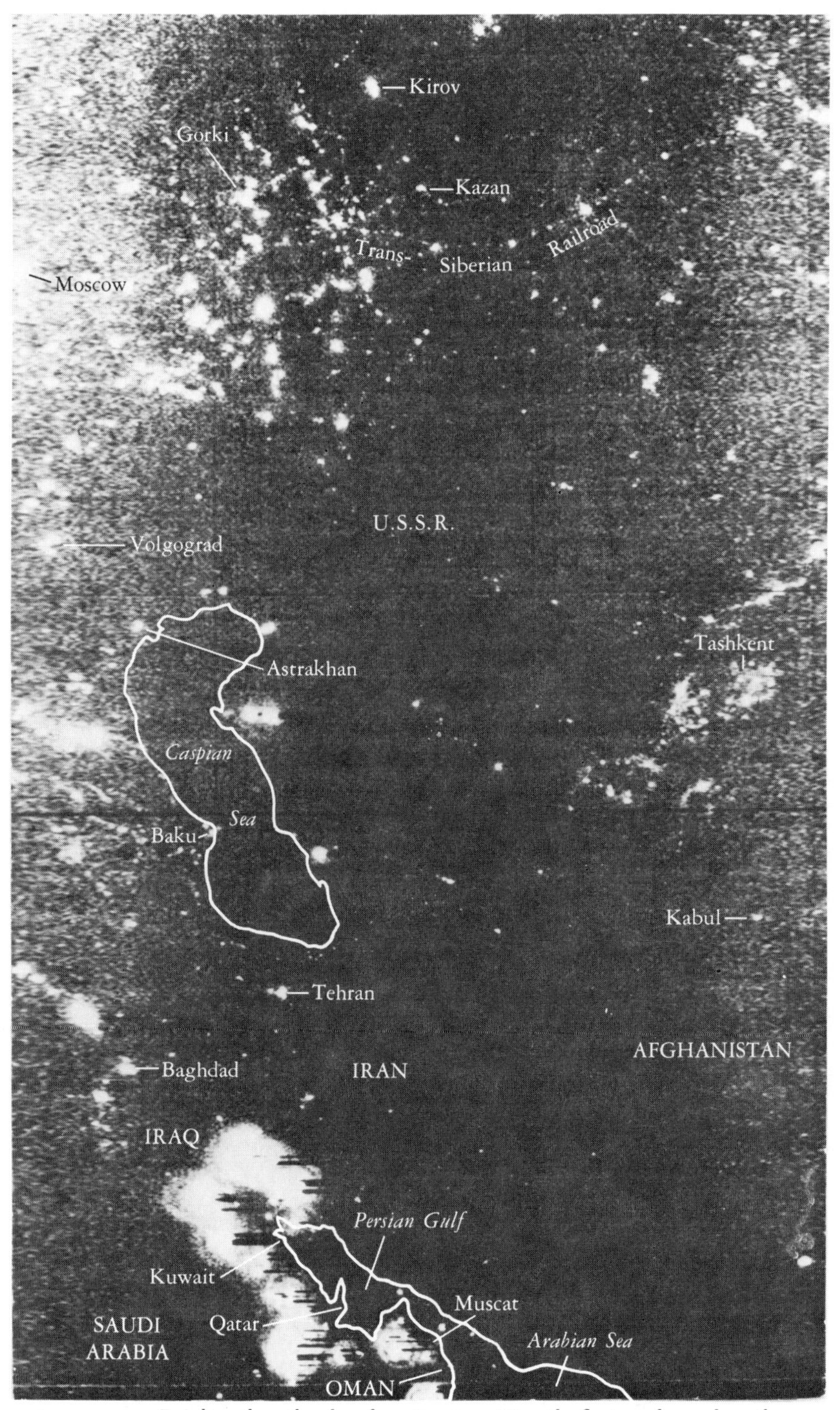

FIGURE 4-3 *Brighter than the glow from Russian cities, the flaring of gas along the Persian Gulf (lower left) lights the night, as shown in this satellite photograph.*

gas, is badly needed. Yet exploration for this fuel is discouraged by the low price ceiling.

The other side of this scarcity is the overabundance of gas in North Africa and the Persian Gulf. In the satellite photograph the flared gas from the Persian Gulf is remarkably bright compared to the lights in Moscow. The gas that happens to accompany the oil obtained from wells is used to satisfy the needs of adjacent countries, but this consumes only a small fraction of the supply. To burn the surplus rather than convert it into fertilizer and other products is a flagrant case of conspicuous waste.

There can be no doubt that the worldwide energy crisis is a terrible reality. But this energy crisis is not simply an energy shortage. The reasons for the crisis are lack of foresight, lack of equilibrium, lack of cooperation, and lack of understanding.

With this book I hope to make a contribution toward relieving this lack of understanding. But it should not be expected that constructive answers are simple. The answers as well as the questions will seem confusing.

Can we hope that confusion will be temporary? Or should we amend the definition of *homo sapiens:* "A tool-using animal who is confused"?

CHAPTER 5

STOPPING THE WASTE

In which we discuss how to save what we lack. The author is reminded of a brief verse from the last century by Wilhelm Busch, originator of the comic strip:

Genügsamkeit ist das Vergnügen
An Dingen Welche wir nicht Kriegen.

An updated and unfaithfully translated version might read:

It is important that we make haste
To save and enjoy what we can't waste.

What must be done is hard to describe and even harder to do.

What is the most inert substance? Is it perhaps the human mind? Or is it possibly the collection of human minds that make up all the thought patterns and traditions of a whole country?

Of course these questions are nonsensical, but they come to my mind when energy conservation is discussed. There is widespread agreement that we waste energy, yet we seem unable to stop the waste. One reason is the inertia of habit. This problem is one of the oldest and most important for human society. How do we change habits gradually and with patience without in our haste destroying more than we build?

But the difficulties inherent in making any change in tradition lie not only in overcoming inertia. The problem of saving a significant amount of energy is intricate. We can imagine a thousand and one ways in which to save energy—some of them excellent, others good, and a few even counterproductive.

Before embarking on this discussion we should review how energy consumption is related to human well-being; obviously there is a relationship. In many parts of Central Africa, where the standard of living is exceedingly low, little more energy is used than that consumed in the form of food.[1] The average rate of energy consumption per person in these countries is probably less than that for all of mankind in the period following the first utilization of fire. If this fact seems surprising, bear in mind that one of the main initial uses of energy was to keep warm—not a problem in Central Africa.

It has been claimed that human well-being (as measured, in a somewhat dubious way, by per capita income) is proportional to the use of energy. Figure 5-1 is a chart purporting to prove this claim. Plotted for the various countries are the per capita income and the per capita energy consumption. It is worthwhile to study these figures to see to what extent energy and "the standard of living" are connected and to what extent they are not.

For a better overview two plots are shown. In Figure 5-1a, we see how per capita energy consumption is related to greater per capita income. (Consumption increases from left to right; increasing income is plotted vertically.) Because of big differences in per capita income, the countries toward the top of the figure are spread far apart. We condense the data in

---

[1]These countries include Guinea, Bissou, Mali, Upper Volta, Niger, Chad, Central African Republic, Uganda, Rwanda, Burundi, Malawi, Ethiopia, and Somalia, which form a belt across the continent somewhat above the equator.

Figure 5-1b, were equal differences on the vertical correspond to equal *percentage* changes in per capita income.

First, it is obvious that no country with a truly low energy consumption (ten million Btu per capita per year or 3 percent of the United States figure) has a per capita income higher than $200 per year. The average citizen of such countries must live in abject poverty. It should be noted that the Central African countries named in footnote 1 do not even appear on the chart. This information indicates that a connection between energy consumption and the good life does exist.

Assuming a general correlation between minimal energy consumption and an intolerably low standard of living, there is another observation we should make. In the third quarter of this century, energy consumption in advanced countries increased twofold and in the underdeveloped world it increased fourfold. In the same period per capita energy consumption increased 1.7 times in the advanced world, but in the developing world it increased threefold. This indicates that the cliché "the rich get richer and the poor get poorer" fortunately does not hold. Energy conservation is possible only in the advanced countries; but it will not suffice. Savings in rich countries cannot offset the continuing and very real needs for more energy in poor countries.

On the other hand, there are also some remarkable indications that energy consumption and the comforts of life are not invariably linked. For example, the countries of western Europe generally have a higher per capita income than the countries of eastern Europe, but not a higher per capita consumption of energy, as shown in the figure. (In this respect, Japan happens to lie in the same class as western Europe.) Specifically, West Germany enjoys a 30 percent higher per capita income than East Germany, but consumes 20 percent less energy per capita.

Often a comparison of the energy consumption in the United States, Switzerland, and Sweden is used to illustrate the wasteful nature of Americans. Indeed, the three countries have similar standards of living, the differences shown in Figure 5-1 having disappeared in the past few years. But Swiss wealth is based on tourism, bank deposits, Swiss watches, chocolate, and cheese. None of these is particularly energy intensive. Therefore, it is not a surprise that the Swiss need about one-third the amount of energy used by the Americans. Unfortunately, the whole world cannot be like Switzerland.

It will be noted that the Swedish use about half as much energy as United States citizens. They use more energy per capita than the Swiss but much less than the Americans. It should be remembered that the strength

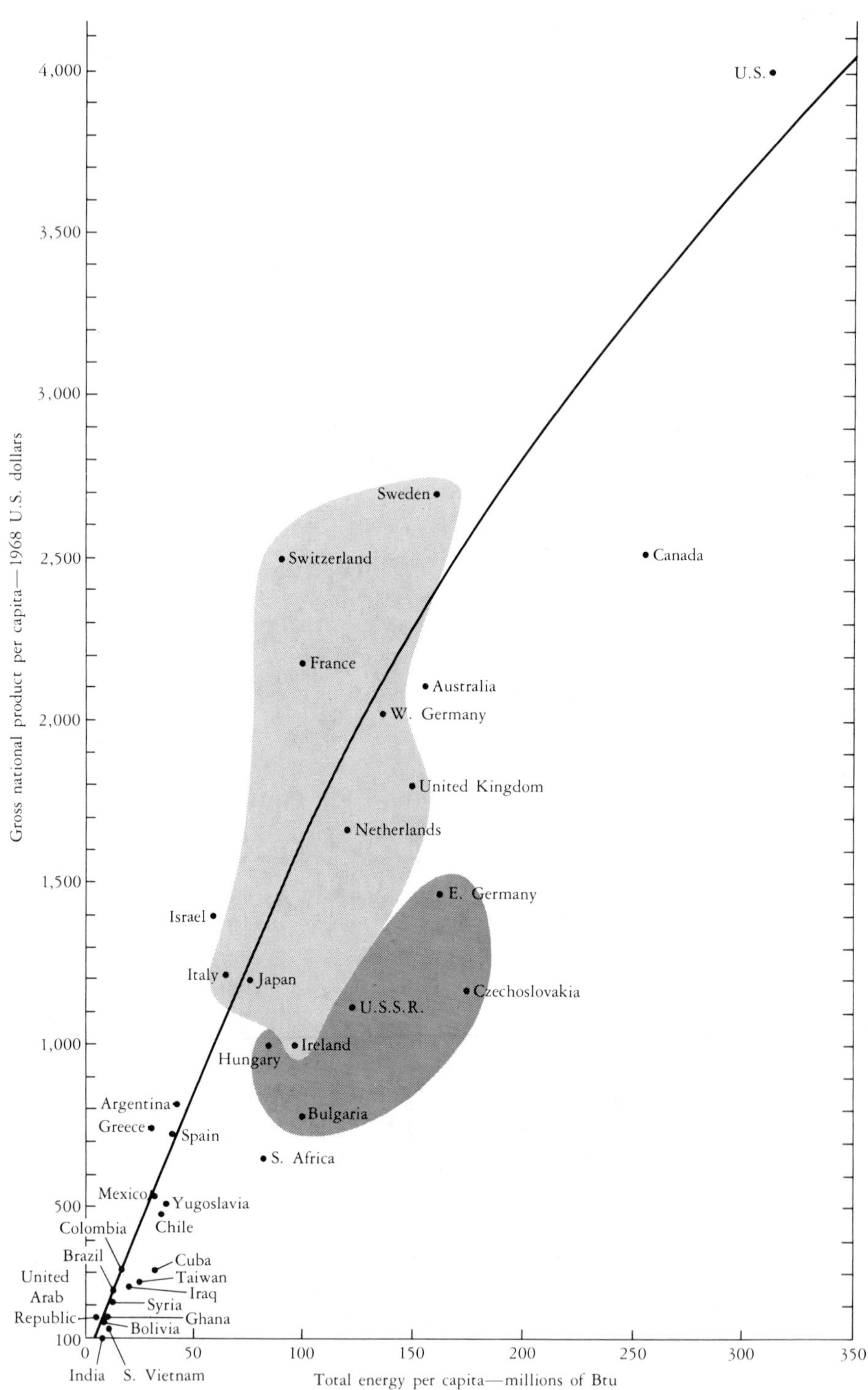

Gross national product per capita—1968 U.S. dollars
Total energy per capita—millions of Btu
4,000
3,500
3,000
2,500
2,000
1,500
1,000
500
100
0
50
100
150
200
250
300
350
U.S.
Canada
Sweden
Switzerland
France
Australia
W. Germany
United Kingdom
Netherlands
E. Germany
Israel
Italy
Japan
Czechoslovakia
U.S.S.R.
Ireland
Hungary
Argentina
Greece
Spain
Bulgaria
S. Africa
Mexico
Yugoslavia
Chile
Colombia
Brazil
Cuba
Taiwan
United
Arab
Republic
Iraq
Syria
Ghana
Bolivia
India
S. Vietnam

FIGURE 5-1 *(a)*

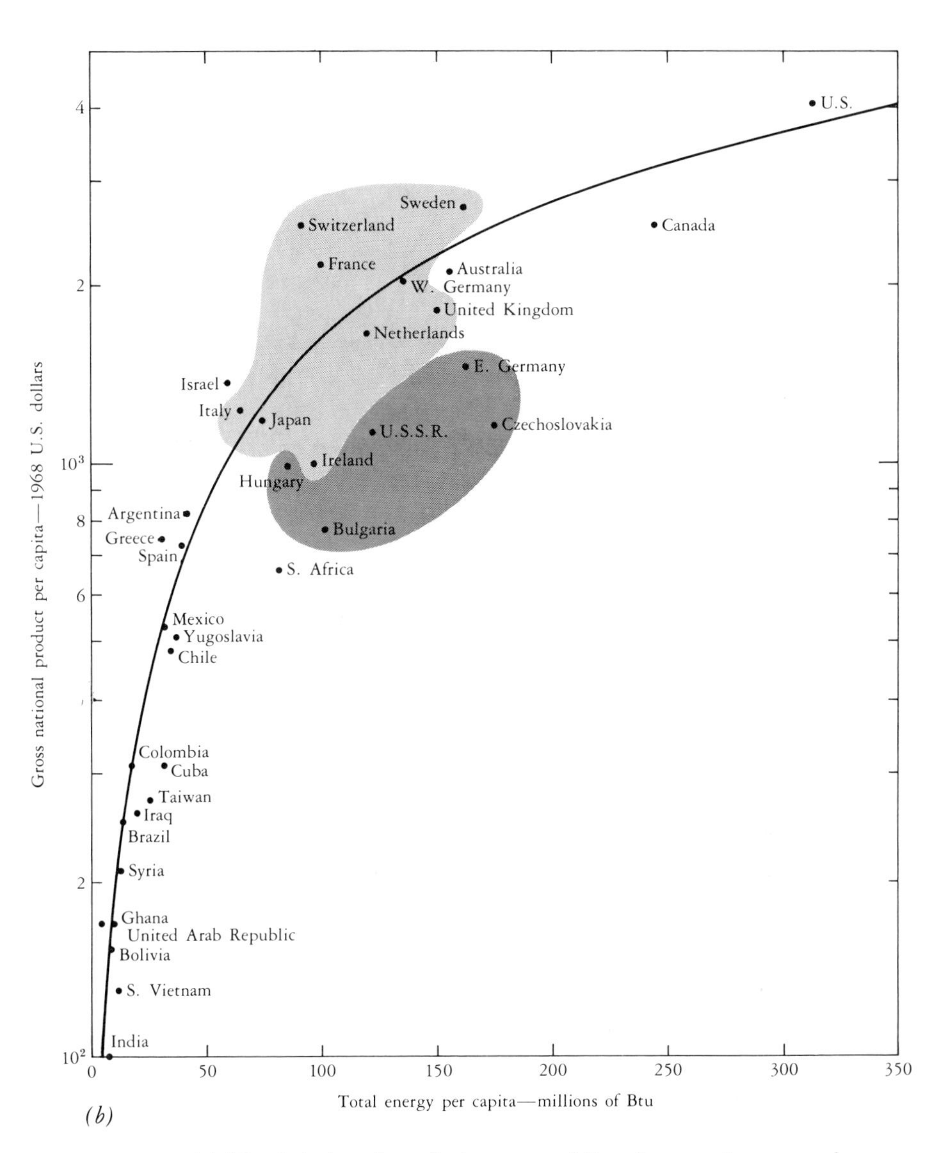

FIGURE 5-1 *(a) The deviations from the heavy curved line show to what extent the connection between the gross national product and energy consumption is violated. Note that a great number of points representing a variety of countries are crowded in the lower left corner. (b) Display of the data from part (a) on a semilogarithmic scale in order to include more, but by no means all, of the countries of low income and low consumption. Note especially the two regions representing eastern Europe and western Europe plus Japan. (Data from* The U.S. Energy Problem, *Vol. 1, ITC Report C645, InterTechnology Corporation, Warrenton, Va., November 1971.)*

of the United States is to a great extent due to its vast size. This provides exceptional opportunities for development, but more energy is needed to overcome the distances. Furthermore, energy intensive agriculture in the United States has provided the international market with three-quarters of its grain supply. It is also true that the smaller energy expenditures in Sweden are related to less waste and reasonable thrift. Here we should indeed learn from Sweden and from others.

How much could we reduce our energy consumption if we did learn? It is probable that we could cut out 10 percent just by acquiring better habits individually. Furthermore, we could rather easily reduce the energy consumption another 10 percent by investing in equipment that is more efficient in its utilization of energy, both for the home and for industry. For homes we should evaluate appliances and air conditioners. Equipment for industry should deal directly with the efficiency of end-use applications of energy, various heat recovery systems, and topping and bottoming cycles, which we shall discuss later.

Some claim that an additional 10 percent could be saved, bringing the total up to an impressive 30 percent. It is real question, however, whether such conservation would require intricate and perhaps self-defeating government regulation. Also, it is possible that energy-saving devices would have to be installed whose construction is expensive and which, in the production phase, may even consume more energy than they are expected to save later.

Let me begin consideration of conservation with an illustration of an individual habit that happens to be quantitatively insignificant but is a good reminder of the need for energy conservation: the matter of lighting. I remember that in Europe as a child I was told that electric lights in daytime are ugly. I distinctly recall that I was puzzled that this should be so. But a strange thing happened. Probably as a result of that early suggestion, I now feel that electric lights in daytime are indeed ugly. Such habits or tastes would be useful to acquire. Whenever I give a lecture on energy I stop at this point and complain that the speaker (meaning myself) is overilluminated. Indeed, with the glaring light I cannot see the audience and I cannot even find out who has fallen asleep. People want to cooperate and turn off the lights, but as a general rule, no one knows where the light switch is. The amount we could save by reducing unnecessary illumination would be relatively small. However, we could effect greater savings by being more modest in our requirements for heating and cooling.

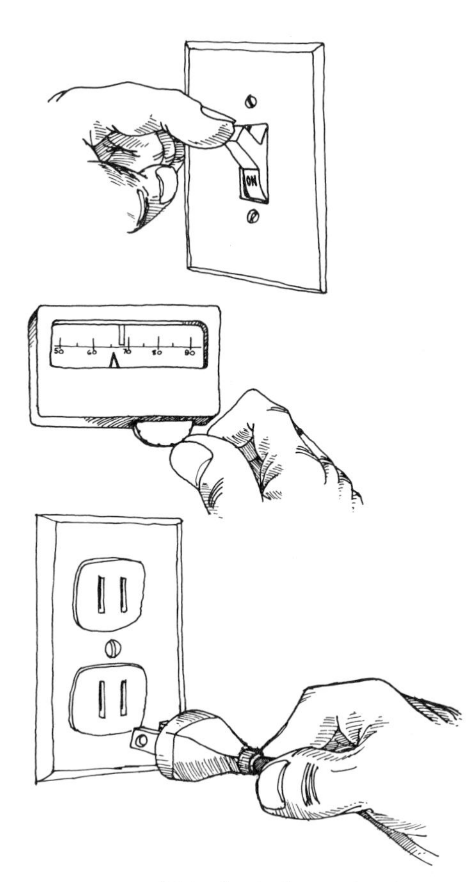

FIGURE 5-2 *This sketch shows what is easy to do. Are we doing it? (George Bing.)*

There is, of course, disagreement on the question of what is an agreeable temperature for rooms in winter and in summer. In Europe rooms are kept cooler in the winter, warmer in the summer. I am inclined to argue that having very warm rooms in the winter and very cool rooms in the summer is the attitude of extremists. It, unhappily enough, is an American attitude. Some of my physician friends tell me that, strictly from the point of view of health, temperature is less important than changes in temperature. To have warmer rooms in summer (by using less air conditioning) and cooler rooms in winter will certainly conserve fuel and might conceivably be better for our health, since we are exposed to less change when leaving or entering a building.

We could further reduce our requirements by wearing sweaters or warm underclothing,[2] and by using electric blankets rather than warmer bedrooms. Many trade-offs are possible. Often it would be more practical to have thermostats and other temperature-regulating equipment in several rooms of the house to adjust the temperature in each as needed. Why should a guest room be heated when no guest is in it?

An even more important case is that of old apartment houses. Utility bills are paid for the whole apartment house and apportioned among tenants or absorbed in the monthly rent payment. Usually there are central controls, or one main thermostat for the entire heating system, and not everyone's preferences can be met. Individual tenants then offset heat deficiencies by using personal heaters in their dwellings; thus additional, less efficient equipment is used and more energy wasted. Even in cases where thermostats are provided for control over individual apartment heat, there is no penalty for overheating in winter or for low temperatures in summer. This is such an important case of energy waste that retrofitting should be recommended and individual apartments separately metered. If one wastes energy, one should pay for it.

All this is boringly obvious. But let me mention a contrast. Think of a warm and friendly fireplace. A part of the feeling of pleasant togetherness it engenders may be that the rest of the house is chilly. On the other hand, let me recall the idea of a man of extraordinary imagination, Buckminster Fuller. He proposed to enclose all of Manhattan in a gigantic dome constructed from polygons, thus creating an artificial climate for millions of people. He had built similar, somewhat less huge domes (called geodesic domes), and they proved quite practical and popular as meeting halls and alternative and interesting forms of housing. Of course, he never constructed a dome on such a gigantic scale as that proposed for Manhattan.

As far as the science of statics is concerned, this plan could be carried out. It is even conceivable that such a dome could be well-insulated. But what about the climate and the pollution within such a micro-universe, which would not be so very "micro" after all? I hate to imagine what could happen in a big earthquake (though careful calculation might justify the plan from this point of view). However, would all those multitudes want

---

[2] When arctic air moved down on us in the winter of 1976–1977, some wore parkas, and not only outdoors.

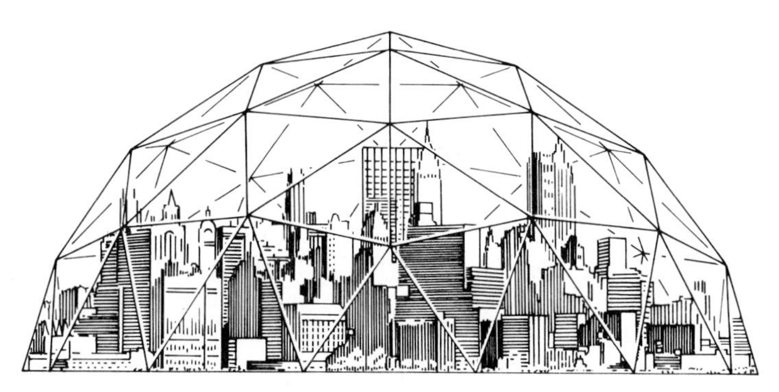

FIGURE 5-3 *Manhattan under a geodesic dome. (Bigger, but better?)*

to live under one cover? I am convinced that our young people would not. In this instance I would join them in asserting that "small is beautiful."

What actually would happen to our dwellings under a reasonable plan of energy economy is probably much simpler than the extremes that have been mentioned: our houses should be better insulated. Most heat is lost through the roof, some through the windows, and quite a bit more through the walls. Methods for better insulation are available. The best and cheapest insulation is accomplished by layers of air that are not allowed to circulate freely. This method employs the same principle as that used in our clothing, quilts, parkas, sleeping bags, and the like. The only difference between insulating our bodies and insulating our houses is that in the first instance we like to use soft substances such as fibers, sometimes further insulated with down; in the case of houses the insulating material should be cheap, like fiber glass or rock wool. Where a double wall with a space is already available, or where there is an attic, retrofitting is easily done by blowing or rolling in the insulation. In windows, double or even triple panes can be used; these are used much more extensively in Europe than in the United States. The use of curtains or other window coverings that will permit solar heat to enter in winter and exclude it in summer is another effective means of conserving energy.

Incidentally, some of the greatest waste today occurs in our mildest climates, where virtually no insulation is utilized. California is the most ultra-American of all American states. California's waste is prodigious, and it is on this sunny yet temperate seacoast that better insulation could improve the situation in the least expensive way. At the time of writing, a $200 tax rebate is offered there to home owners who install reasonable insulation in their houses. In the long run, construction of new houses with good insulation will be of greater importance, since in this case energy-saving can be offered at little added expenditure.

In principle, big savings could be achieved by use of the "heat pump." This is really an air conditioner run in reverse. What the air conditioner accomplishes is to cool the interior of the house but (since energy is conserved) heat the exterior. You can check on this statement on any hot summer day if you walk past the windows of an air-conditioned motel where the older type "window box" system is used. With the heat pump, by contrast, the cold air outside is further cooled while warm air is blown into the house. There is a lot of heat even in cold air (as will be discussed later in connection with the second law of thermodynamics), and the heat pump is so constructed that the least energy should be needed to maintain the desired temperature difference. It turns out that use of this appliance leads to considerable savings where the temperature difference is small. On a really cold day with a temperature difference of several tens of degrees a heat pump uses less energy than would be used in direct heating. But the saving is not very great and the heat pump uses electricity, which as a general rule should be considered the most expensive form of energy. Heat pumps therefore may not be the appropriate solution where winters are harsh.

What we are discussing is really a combination of improved habits and improved technology. The old American traditions of privacy and independence would be preserved in this way. Individual houses are still preferred, but in a duplex or a bigger cluster of units there is less exposed wall surface, with correspondingly less heat loss in winter and less need for air-conditioning in summer. In keeping with this advantage, apartment houses or the slightly different and more pleasing arrangement of "condominiums" are increasingly popular.

A far more radical approach would be to build dwelling places underground. Of course, this plan would require more lighting and much more ventilating equipment. In climates with cold winters and hot summers the savings on heating and air-conditioning would much more than offset the cost of lighting and ventilation. The question of moisture in the earth would have to be dealt with, and probably such construction would not be feasible except where the dwelling place could be constructed above the water table. It took mankind a long time to get out of caves. It may seem particularly objectionable to get back into anything that resembles a cave, no matter how well decorated and engineered it may be. Some people have claustrophobia and would never be comfortable in such a home. Of course, during the Ice Age a large proportion of all people lived in caves.

A particularly interesting development is that of "mobile homes." These are manufactured by a relatively small number of enterprises. As a general rule they are not truly mobile. The wheels are used only to move the structure to its eventual location where the mobile home becomes sedentary. These dwellings are a relatively inexpensive, satisfactory, and widely used form of housing. Almost half of the newly sold single residential units today are mobile homes. A few years ago insulation for them was insufficient, sometimes nonexistent. Insulation is now available for mobile houses at relatively little added cost, and some states are making adequate insulation mandatory through legislation. The conventional home-building industry, with its greater inertia, is following suit at a somewhat slower rate.

Transportation presents problems even more serious than those of housing. Nowhere but in the United States could the automobile industry have developed so extensively as it has. Our country's great spaces provided the need, desire for privacy provided the motivation, and development of mass production methods gave the opportunity. From all these factors there followed universal adoption of a vehicle that permits the population to spread out and yet maintain contact with places of work and "neighbors" within many miles.

Now we are stuck with our cars, for good and for ill. The automobile consumes about 20 percent of the energy produced in the United States. It is the leading single user of that fuel whose shortage is most acute—petroleum.

This situation poses a sensitive problem. It will be solved only if technical improvement is accompanied by public willingness to change. Two contributions will be required: first, some slight changes in lifestyle, and second, the redefining of our transportation concepts and customs.

One alternative may be rapid mass transportation, and in some areas this is a viable alternative. But in much of the United States the housing is distributed at some distance from working places, making the car a real necessity.

In New York rapid transit works—in fact, it is indispensable—though it has to be subsidized. In Washington, D.C., a subway system was recently started, and it seems to serve a useful function. In the San Francisco Bay Area, BART has been advertised as the latest that the space age can offer. University of California students have nicknamed shuttle

FIGURE 5-4 *The Environmental Impact Statement is missing. (George Bing.)*

buses that carry them to BART stations "Humphrey GoBART." In most respects the system seems to be disappointing. BART is expensive and manages to carry only a small percentage of the commuter traffic. In Denver a rapid transit system has been planned, at a projected cost of $5 billion, for completion in the year 2000. Its advocates claim that it will accommodate 15 percent of the passenger mileage. While rapid transit systems make a limited contribution, as a nationwide solution they are impractical. Unless the whole country is to be rebuilt we must manage to get the best possible use from the automobile.

The use of legs for locomotion is an art that was almost lost but is now reviving in the enthusiasm for walking, bicycling, and jogging. (I wonder if people could do their jogging as they go to work; but this might be considered too radical a change in lifestyle.) For those who are lazy or who want to go a little faster, a compromise is recommended. On the "moped," a bicycle with a tiny motor attached, the rider can propel himself either by the motor (hence "mo") or by pedaling ("ped"). The moped may have a top speed of 20, 25, or 30 miles per hour, weigh around one hundred pounds, and average from 130 to 220 miles per gallon of gasoline. These vehicles can be found throughout Europe, Southeast Asia, and Bermuda. In France, one out of nine inhabitants uses a moped as the basic means of transportation. But until 1974 mopeds were unavailable and technically illegal in the United States, not because they are unsafe but because they did not fit into any convenient niche in our motor vehicle laws.

At the time of the 1973 oil embargo, carpools were established. Highways became less congested. But now, a few years later, we seem to have slipped back into the inefficient old habits.

Improvement in car construction is even more important. Mileage per gallon can be greatly increased. The 1978 General Motors fleet average is 18.6 miles per gallon, compared with 12.3 miles per gallon in 1974. Much of this improvement was accomplished through design changes and weight reduction through increased use of aluminum, and with no sacrifice of interior space or comfort.

Another important, ingenious, and effective means of fuel saving in automobiles resulted from the introduction of electronic controls. The efficiency of an internal combustion engine is never high. But it becomes particularly low if the gasoline-air mixture is not optimal and if the mixture is not detonated by the sparkplug at the right moment. All these factors are adjustable. The optimum shifts considerably, however, depending on whether the car is idling, accelerated, or driven at a steady speed. The electronic control performs the adjustment almost instantaneously for the best use of fuel at all engine speeds.

An earlier advancement was the use of radial tires. With rubber strands simply circling its circumference, the radial tire is more elastic and converts less energy into heat than do older types of tires, which use crossed strands set to angle acutely from the direction of the tire's roll. A tire is flattened and distorted as it rolls on the road. With crossed strands the angle of crossing changes, resulting in surprisingly great friction and energy waste.

Gasoline consumption rose with the introduction of passenger conveniences engineered with maximum regard for style, comfort, and reliability but small regard for fuel economy. To note an extreme example, it is particularly inefficient to equip a dark-topped car with a powerful air conditioner to compensate for excess heat that is absorbed because of the car's color.

Too often equipment aimed at pollution control has been constructed without regard to energy economy or without adequate consideration given to the precise nature of the pollution. Changes in emission-control devices made a few years ago to eliminate unburned hydrocarbons by the use of catalysts accomplished that purpose, but they also increased carbon monoxide emission. After extensive discussion a tentative conclusion was reached, that the result was a "balanced" situation. The damage caused was about the same as the advantage gained, disregarding the expense of the equipment.

Modern stratified charge cars, first introduced by the Japanese, are relatively clean and efficient. The burning process occurs in two strata.

First, fuel mixed with a little air is injected into a cylinder where it burns at high temperature. Then more air is added and the burning is completed at a lower temperature. This leads to better use of fuel and fewer polluting products in the exhaust.

Approximately twice as efficient as gasoline engines are diesel engines, which burn droplets of the fuel in air. At present three European manufacturers have representatives of these cost-efficient passenger cars on the American market: Peugeot from France and Volkswagen from Germany for the economy-minded, and Mercedes (also German) in the more luxurious, status-conscious category. All three have been received widely and well. Among them Volkswagen is the newest entry to the scene and ranks highest in fuel economy, while Peugeot and especially Mercedes have represented the diesel market for many years. General Motors now has diesel passenger cars among their 1978 models. Diesel fuel has been relatively scarce and hard to find; not every corner gasoline station sells it. But this trend is changing. It is increasingly available and better publicized. Prices for diesel fuel have increased noticeably as well with demand. It used to be half the price of regular gasoline but is now only a few cents less per gallon. Even so, it is still much cheaper per mile.

How can all the desirable transportation changes be brought about rapidly? One obvious possibility is to double or triple the price of gasoline by imposing a heavy tax. This would raise American gas prices to the neighborhood of European prices. Americans, however, are much more dependent upon driving; distances between their homes and working places are, as a rule, greater than in Europe. What is acceptable in Europe would seem to be a great hardship in the United States. This example illustrates how tradition can make desirable changes difficult to effect. Europeans get the impression from our low gas prices and high consumption that we are behaving in a truly wasteful manner. Thus it is harder to establish genuine understanding between western Europe and ourselves concerning energy.

Other possibilities have to be considered. High taxes could be imposed on cars that use a great deal of gasoline. Small trucks and taxis could be equipped with diesel engines. Since these vehicles are driven many miles in a single year, their turnover is rapid. The introduction of diesel engines could therefore have an early and highly desirable effect. Such a measure might be enforced for new cars by regulation. Big trucks already have diesel engines.

Improving the energy efficiency of equipment for general public use is, clearly, a technical problem. It is up to the public to buy such equipment, however, and even to demand it. Generally a less efficient appliance is less expensive, conveying a false sense of economy. The customer must ultimately pay more because of inefficiency in energy. It is a case that might be described as "buy cheaply now, pay later to the Arabs." For instance, an air-conditioning unit selling for $120 retail delivers 4,580 Btu of cooling per kilowatt hour. For $165 you can buy a unit that delivers 8,700 Btu per kilowatt hour. This means an 89 percent improvement in efficiency at a 30 percent increase in cost.[3] Of immediate interest to the consumer is how much money he will save per year on operation of the unit. In Boston's climate the unit would work five hundred hours per year, and the consumer would save $18 per year. In a little over two years his additional $45 investment would be paid back. It seems desirable to label energy-consuming equipment to show not only the sale price but the probable cost of operation. This, of course, would vary with location.

There is another, perhaps more impressive, view of the same situation. Inefficient air conditioners require the installation of additional generating equipment, and thus a greater capital investment. It turns out that use of the better air conditioner, costing the customer $45 more initially, will not only repay his investment but will effect a $200 savings to the utility that provides electricity, by making it unnecessary to install added generating equipment.

The conclusion is not so clear as it appears, however. New England today has a 40 percent excess in generating equipment, and therefore will not need to install added sources of electricity for some years to come, even if the public buys wasteful air conditioners. What should be done is not quite clear, but that the buyer should at least be informed about the consequences of his choice is obvious.

One reason for the high energy consumption in the United States is our industrial society's practice of rapid obsolescence.[4] This has been a consequence of mass production and of continually improved products. It has appeared desirable to replace old equipment rather than repair it, and American industry has been geared to that. Even apart from the fact that all industry runs on energy, many materials that were obtained by great expenditure of energy are thrown away. A good example is aluminum.

---

[3]Elias P. Gyftopoulos and Thomas F. Widmer, "Energy Conservation and a Healthy Economy," *Technology Review*, 79, no. 7, June 1977.

[4]"Ending is better than mending" is the chant of learners in *Brave New World* (Aldous Huxley, 1932) as they are "adapting future demand to future industrial supply."

Aluminum-rich minerals are common and cheap. Aluminum's comparatively high cost is due to the heavy amounts of electricity needed to convert those materials to metallic form.[5]

This topic is obviously related to ways in which industry can contribute to saving energy. The topic is difficult, not because there are inherent scientific or technical obstacles, but because of the immense force of tradition and the variety of ways in which this tradition could and should be changed. Throughout its whole history American industry, while saving human labor, has squandered energy. In the present energy crisis we are apt to forget how immensely important labor-saving devices have been and still are. But in the last quarter of a century those understandable trends have become exaggerated by the reaction of economic forces. Energy has become ever cheaper; at the same time, in a healthy economy with practically full employment, human labor has become ever more expensive. The resulting energy waste has been unavoidable. It has permeated every nook, cranny, and invention of our industrial machinery, and is deeply imbedded in the tradition of how to run an industry. Changes, which in the past few years have become obviously necessary, require a change in traditional procedures and therefore will be difficult to accomplish.

What appears to be an obvious answer is the wrong one. Let us *not* throw away the old equipment, which literally devours energy. A massive replacement program would be too expensive, both in money and in energy. The general rule that must be heeded is to find a reasonable compromise on a case-to-case basis. When, as here, rapid change is indicated, good ideas will be particularly valuable, especially about details. What we can propose here are only generalities. What is really needed is work that would have to be performed by a million gifted engineers.[6]

One way we save energy is to recycle materials from obsolete equipment. Recycling has the further advantage of providing a way to dispose of trash while saving valuable material. Coors Brewery buys back used aluminum cans, which gives them access to cheaper aluminum through

---

[5] Pound for pound, aluminum requires between ten and twelve times as much energy as iron. In most uses one pound of aluminum will replace two or three pounds of iron. Therefore, to replace iron with aluminum may cost four to five times as much. If instead of starting from a mineral one starts from scrap aluminum, there is a twelvefold reduction in the energy cost of fabrication.

[6] Please note that these engineers work as individuals. If they worked as a committee, the inertia of one million human brains would completely immobilize them.

saved electric current. The practice also diminishes the number of beer cans adorning landscapes in the Western states.

Perhaps we should favor a more radical approach. Though reprocessing is good, prevention of premature obsolescence is better. Washing machines should not have to be scrapped because one small part cannot be replaced. Today there is not much choice. Spare parts are not generally available for old equipment, and repair has a hard time competing with replacement. The practice of buying new is deeply imbedded in the habits of Americans and in the way our industry operates. There are few occasions indeed where I would advocate a regulatory approach, but there should be a means, by regulation if necessary, to make the availability of spare parts and the maintenance of old equipment the rule rather than the exception.

Turning to technical efficiency and energy conservation in industry, we consider an impressive example of the energy savings obtainable by using current technology: *cogeneration,* that is, the combined production of electricity and industrial process steam. Use of cogeneration means the reduction of one type of energy waste; electricity produced by cogeneration uses fuel at least twice as effectively as that achieved by the most efficient central-station electric generating plant. In West Germany cogeneration accounts for approximately 18 percent of electrical needs, compared to 10 percent in Italy and only about 5 percent in the United States. If the United States had achieved the same level of cogeneration as Germany, our fuel savings in 1975 would have been more than a quad per year, or the equivalent of half a million barrels of petroleum per day. A recent study by Thermo Electron Corporation for the Federal Energy Administration claimed that in just three industries—papermaking, chemicals, and petroleum refining—there exists the opportunity to produce over 34 percent of all the nation's electricity by means of cogeneration and waste heat recovery, a figure never approached by any industrial society. Let me quote the analysts:

> While long-term dramatic improvements in end-use efficiencies can probably be made throughout the economy there will be a significant capital cost involved, unlike the case with many of the simple measures already implemented in response to rising energy prices. Such conservation actions, involving the trade-off of energy cost savings against initial capital costs, deserve the most careful attention in formulating a new U.S. energy policy.[7]

---

[7]Gyftopoulos and Widmer, "Energy Conservation and a Healthy Economy."

What has been estimated as the potential contribution by cogeneration may be too high and, indeed, may have to be reduced considerably when individual problems of technical execution must be faced. However, it appears certain that such processes could make substantial contributions toward an efficient energy economy.

In order to understand how a process like cogeneration works in an ideal way, we must discuss some of the laws of *thermodynamics,* the science of heat and its exploitation.

The first law of the science of thermodynamics is called "the conservation of energy." Accordingly, there is no *energy* shortage. In fact there can never be an energy shortage. By the laws of nature, *energy is conserved.* We can neither lose it nor generate it. When we speak of a shortage of energy, we mean energy in a usable form. Mechanical work is usable. Available work is the essential basis for any rational assessment of efficiency. Electrical energy is, in a practical sense, equivalent to mechanical energy and is equally usable. Hot bodies have energy content; so have cold bodies, though their energy content is less. But heat energy in itself is not usable. It can be used only if we have a temperature *difference.* The greater the difference, the more usable the heat energy.

The maximum usable energy in any given situation was discussed in 1824 by the French physicist N. L. Sadi Carnot. A remarkable point in his accomplishment was that he formulated the essential statements about energy efficiency before the concept of the energy content of heat had been clearly defined. (This may be a lesson to those who imagine that the progress of science follows a logical course.) *Carnot's principle,* that the efficiency of a reversible engine depends on the temperatures between which it works, led to the formulation of the second law of thermodynamics.

To understand Carnot's rather simple rules we should discuss the way in which temperature is measured. A scale now obsolete, though still in general use in the United States, was introduced in 1714 by G. D. Fahrenheit, a German instrument maker, in Koenigsberg, East Prussia. He chose as 100° the normal human body temperature and 0° as the lowest temperature that he could observe (that is, the lowest in Koenigsberg; he should have been around in Chicago during the early months of 1977). A better defined temperature scale is the centigrade, or Celsius, scale, named for Swedish astronomer Anders Celsius. On this scale, 0° temperature is defined as the freezing point of water, and 100° as its boiling point at normal atmospheric pressure.

The thermodynamic temperature scale was proposed by William

Thompson (Lord Kelvin) in 1848.[8] It is perhaps immodestly called the absolute temperature scale because it is independent of the properties of any particular substance used in a thermometer. Recognizing that temperature is disordered motion, Kelvin designated the point where all motion due to temperature comes to rest as 0° K (read: zero degrees Kelvin). Indeed one cannot get any lower. This scale is simply calculated from the centigrade scale by adding the number 273.1. Thus on the Kelvin scale water freezes at 273.1° K, and normally boils at 373.1° K, keeping a 100° interval between ice and steam. The temperature on this scale is designated by T.

Using the Kelvin scale, Carnot's efficiency—the maximum efficiency of an engine—can be written in simple fashion. If two temperatures are available which we designate as T_{high} and T_{low}, then the maximum useful work obtainable is

$$\frac{T_{\text{high}} - T_{\text{low}}}{T_{\text{high}}}$$

That is, we take the temperature difference and divide it by the higher of the two temperatures. Obviously the efficiency never can be greater than one. For an efficient thermal engine, therefore, we need high initial temperatures. The low temperature in the equation is practically limited to the lowest temperature in the surroundings.

A general rule is the greater the temperature difference, the greater the expense. A hot flame requires valuable fuel. Small temperature differences are available cheaply in the form of waste heat or solar heating using flat plate collectors. Industry accounts for 40 percent of our fuel consumption. Of that amount, 40 percent is used as process steam, a convenient medium for making heat available. Thus, 16 percent of all our fuel goes into process steam. Most of the energy needed for process steam is required in converting water into steam. This first step in the production of process steam needs a relatively small temperature difference, particularly if the water is boiled under slightly reduced pressure. Once this relatively cool steam is produced, heating it more requires little energy, but energy of higher quality. One could use a flame or simply compress the steam,

---

[8]In 1851 Lord Kelvin stated the principle of the dissipation of energy, briefly summarized in the second law of thermodynamics. It was he who, in 1848, recognized and pointed out the full value of Carnot's principle of the reversible cycle.

raising its temperature but using expensive mechanical energy. The bulk of the energy, cheaply delivered evaporation, could be combined with a smaller quantity of heat energy from a more expensive source. The cost of process steam could thereby be cut in half.

It is probably even more important to use process steam for more than one purpose. The steam could serve industry and also generate electricity. It is here that we can make use of the Carnot efficiencies, or to be slightly more highbrow, the second law of thermodynamics. This point can be illustrated with two specific and opposite examples.

The first is a case where high temperature process steam is needed. After the steam has served its purpose and is still quite hot, it is released into the atmosphere, and at this point it is called, not quite rightly, thermal pollution. Instead, this steam could be used in what is called a *bottoming cycle*—methods designed to exploit the lowest temperature at which energy can be extracted. In principle the steam could be used to drive a turbine that would then generate electricity. One difficulty is that the temperature of the steam is not very high, and consequently energy can be generated with only moderate efficiency. This is unavoidable. But we also find a second and avoidable problem: the steam, which is not very hot, has a low density and a big volume, so big turbines would be required, at great capital investment. It would be better to transfer the heat to another *working fluid,* for instance a freon-type substance similar to the fluids used in refrigerators. Such a substance boils at a lower temperature than water and its vapor has higher density. The result is to use relatively cool freon steam to generate electricity in smaller equipment and at a more reasonable capital cost. In some instances this procedure is already in use.

The opposite extreme is the case where low temperature process steam is needed. Here a little more fuel may be used so that steam is produced at a temperature considerably higher than is needed for the industrial process. Electricity could then be produced from the high-temperature vapor, and the process steam would be obtained from the exhaust of the electric generator.

As mentioned earlier, processes of this kind are called, for obvious reasons, cogeneration of electricity. Cogeneration has the advantage of needing not only less fuel for a given amount of electricity produced, but also less capital investment. Producing steam requires machinery which, in the case of cogeneration, serves two different purposes.

The Carnot efficiencies are important not only in cogeneration but in all plants generating electricity. It is correctly said that the average efficiency of generating electricity today is approximately one-third. According to the laws of thermodynamics it could hardly be 100 percent; one cannot start with fuels at temperatures too high, nor end with temperatures too low. On the other hand, hydroelectric power approaches the ideal of 100 percent efficiency.[9]

One obvious side effect of low efficiency in generating electricity is the appearance of waste heat, which, as stated, many call thermal pollution. A temperature increase in a body of water may in fact have advantages as well as disadvantages; public opinion at present seems concentrated on the disadvantages. Raising efficiency from one-third to two-thirds is difficult but not impossible. If that were done, waste heat from electricity generation would fall from two to one to a value of one-half to one. Waste heat per unit of electricity thus would be decreased by a factor of four.

Today's one-third average efficiency is due to the fact that new generating plants reach an efficiency above 40 percent, while some old plants are well under 30 percent. Higher averages will be reached in the future not only by retiring old plants but also by introducing new methods. As long as steam turbines are used exclusively for driving electric generators, it will be expensive to obtain power from low temperature steam.

Bottoming cycles can be used to improve efficiency. By adding the equipment needed to transfer heat to freon or some similar substance, efficiency can be increased from 40 percent to 50 percent, at an optimistic estimate. This technology exists but is not widely used as yet.

Similar and possibly bigger gains can be independently achieved in *topping cycles,* in which the top of the available temperature range is exploited. Typical generating equipment does not use the high temperatures that flames produce because it is difficult to find materials that withstand them. This is particularly true when water vapor is around; at high temperatures it will attack any metal and many other substances.

[9] In accounting for electrical energy the figures used are not the electrical energy delivered but the amount of heat energy required to produce the current. This energy is, on the average, three times greater than the energy delivered in the form of electricity. The convention is used even in connection with hydroelectric power. Though obviously no fuel is consumed, the energy appearing in a statistical survey is the energy that would have been needed had we substituted oil or coal plants for hydroelectric power. The problem of accounting will, of course, be further complicated as we make progress with cogeneration. The preferred methods that are in general use will also appear in a summary of energy expenditures in Chapter 13.

Nevertheless, attempts have been made to construct complicated, clumsy-looking turbines from "cermets," hybrids between ceramic and metallic materials, or other heat-resistant, moisture-resistant substances. Limitations generally are not in reaching the high temperature but in keeping the apparatus operating under those violent conditions.

Exotic proposals have been made that hold some hope for meeting the problem. One possibility is *thermionics.* Electrons are boiled off very hot tungsten plates and condensed on a colder tungsten plate. Tungsten resists heat in an excellent fashion provided one is selective with other chemicals that are in contact with it. In thermionics only one additional element, caesium, need be present, and that in small amounts. Caesium's virtue is that it clings only weakly to its outermost electrons, and it therefore is a good material to use in a substance called *plasma.* Consisting of electrons and positive ions, plasma is the substance that must be present between the tungsten plates and that allows easy passage of electrons. (Plasmas are quite common in nature; the interior of the stars that contain most matter in the universe can be described as plasma.[10]) Thermionic generators can give up to 20 percent efficiency and yield temperatures high enough to be utilized in the most effective power plants. But lots more work will be needed before reliability, mass production, and low costs can be established for thermionic topping cycles.

Another topping cycle that appears difficult but promising is the MHD Generator, a happily abbreviated name for the Magneto Hydro Dynamic generator. In this generator caesium or potassium, a cheaper substitute, is added to the hot combustion products. The gas is partially ionized, that is, some electrons and positive ions appear. High pressure drives the ionized gas down a channel. Across this channel is maintained a magnetic field that deflects the positive ions to the right and the electrons to the left. These are collected on opposite sides of the channel's surface, and can drive an external electric current. The beauty of the procedure, as with the thermionic generator, is that there are no moving parts, only flowing gases, plasmas, and currents. The caesium or potassium is recaptured and reused. The remaining burned products go on to drive the normal state of electric generating equipment, such as a steam engine. The main difficulty is the behavior of the materials in the channel in contact with the burned products.

---

[10]We shall return to the discussion of plasmas in Chapter 11.

If the financial problems of bottoming cycles and the severe technical problems of topping cycles could be solved, a two-thirds generating efficiency could be attained. One might imagine accomplishing even more. A future electric generating plant with a high initial temperature, T_1, and a final discharge temperature T_4 may consist of a topping cycle working between T_1 and T_2, a normal cycle between T_2 and T_3, and a bottoming cycle between T_3 and T_4. The topping cycle works with the efficiency $(T_1 - T_2)/T_1$, leaving a fraction of the original energy, T_2/T_1, for the normal cycle. This is utilized with the efficiency of $(T_2 - T_3)/T_2$, or relative to the original energy $(T_2 - T_3)/T_1$, leaving for the bottoming cycle a fraction T_3/T_1. This in turn is used with the efficiency $(T_3 - T_4)/T_3$, or, relative to the original energy, $(T_3 - T_4)/T_1$. Total efficiency relative to the original energy available is

$$\frac{T_1 - T_2}{T_1} + \frac{T_2 - T_3}{T_1} + \frac{T_3 - T_4}{T_1} = \frac{T_1 - T_4}{T_1}$$

a result we could have guessed. Since T_1 may well be 3000° K, and T_4 could be 300° K, an efficiency as high as 90 percent seems possible. Unfortunately, we shall never attain these ideal efficiencies. There are engineering losses in each stage and further losses in the transfer of energy between successive stages. An efficiency of two-thirds is as high as we can expect from practical electric generators.

There exists, however, a possibility of almost 100 percent efficiency for the conversion of fuels into electricity. The idea is a hundred years old and it is quite straightforward. The chemical energy derived from the rearrangement of the atoms in gasoline and in oxygen, for instance, should never be permitted to produce a high temperature. Instead, the rearrangement should proceed through an intermediate state, an aqueous or other solution called an *electrolyte* containing positive and negative ions—that is, atoms or groups of atoms with a deficiency or excess of electrons. Electrolytes are as common as sea water and as useful as the fluid in batteries. The trick is to pursuade chemical reagents to enter the solution. If the reagent is hydrogen, this can be done with the help of platinum, which takes up hydrogen molecules and can deliver them into water as hydrogen ions. The efficiency of fuel cells is not limited in an obvious way by Carnot's principle or its formulation in the second law of thermodynamics because no temperature changes need be produced. Since charged particles can swim around in water, the possibility is open for a direct conversion of chemical into electrical energy. Such efficient fuel cells have been produced and work reliably in the American space program. Unfortunately, they are still much too expensive for general use. No way has been presented to make fuel cells cheap enough for general application in the forseeable future.

Doubling the efficiency of electrical generation could be the single biggest achievement in energy conservation. The tools of our most advanced technology have been deployed for this purpose. But numerous easier methods, such as cogeneration, are available today.

Other industrial patterns should be considered in seeking energy conservation. Many industrial processes require that water be added and subsequently evaporated, with the latter step of course consuming energy. One example is the paper industry. Wood fibers and water make pulp; dried pulp is paper (this is a vastly oversimplified explanation of the process). The less water was used originally, the less heat is needed eventually. It is certain that a great portion of the heat can be saved.

Something similar happens in the cement industry. Adding water causes the pretreated rock to be easily molded. Energy is then required to remove the water. European and Japanese cement processes, however, use less water and less energy.

It is even simpler (in principle though not always in practice) to have furnaces with better heat insulation and arrangements to recapture and use the waste heat in industrial processes. What is most often applied is not most economical.

There is a practice of hardening surfaces of tools by letting foreign atoms penetrate these surfaces to a shallow depth. As a general rule, this process requires high temperatures to propel the foreign atoms, for example carbon atoms, and to loosen the surface so that the foreign atoms can enter. It might be better to leave the surface at normal temperature and shoot the foreign atoms into it at a high speed, which can be attained if these atoms are ionized and accelerated by an electric field.

In most industrial processes a theoretical value is given for the minimum fuel requirements to accomplish the needed results. The value may depend on the energy that must be added, on the Carnot efficiencies, or on other circumstances. It has been estimated that actual utilization of energy in American industry is only 8 percent of what it could be. The further guess was made that this could be increased to 33 percent, a great improvement if it could be realized.

The *if* is a big one. Of necessity it involves a persistent comparison of apples and oranges. Minimizing energy input is not the main purpose of industry. In fact, practically no attention was paid to energy consumption until recently. Nevertheless, it makes sense to compare what is done with what could be accomplished within the laws of physics.

One practical criterion has been proposed: balancing energy conservation against added energy production. In many of the cases mentioned above, energy-saving devices require capital investment. We may well ask which costs more in investment: to produce an additional barrel of oil a day or to save an additional barrel a day. In making such an estimate, however, it is important to include all relevant costs. It costs money to search for oil, for instance, and to put oil pumps in place, transport the oil, distribute it, and pay for any adverse environmental impact. On the other hand, with equipment designed to save energy, one should account for not only the cost of that equipment but the effects of delays and discomfort connected with the changeover. Drawing such a balance will be to some extent arbitrary. In order to envision the amounts of money in question we can remember that the production of a barrel of oil a day (or its equivalent in some other form of energy) requires an investment of between $10,000 and $15,000. If we want to consider the big units of quads per year, we would be talking about an investment of between $4.5 billion and $7 billion to produce this additional supply of energy. It should be noted that the capital cost of electric generating units is much higher still. An electric plant and electrical distribution system that consumed one quad per year in fuel would cost between $15 billion and $20 billion.

An approximation of the distribution of potential energy savings and investment levels in certain industries, as shown in Table 5-1, was recently determined. If the estimates in the table are right, then approximately $11 billion are needed to save a quad per year. In my opinion, the conclusion to be drawn from this table is both interesting and uncertain. On the one hand the cost of conservation seems reasonable. On the other hand, while new equipment is required for conservation, new methods of energy production may well appear and make this attempt at conservation uneconomical. At any rate, independence from OPEC oil sources is certainly an important objective.

The practical answer may be connected to the percentage saved. To try and save as much as one-fourth may be too costly. If only the top 10

TABLE 5-1 *Approximate Distribution of Potential Energy Savings and Capital Investments for Selected Industries*

Industry	*1985 Quads saved (at equivalence*)*	*Energy saved (percent)*	*Capital required ($ billions)*
Steel	0.77	17	8.4
Paper	0.80	26	8.7
Aluminum	0.14	15	1.5
Cement	0.26	42	2.8
	1.97 Quads	24% (average)	$21.4 Billion
Overall Industrial Sector	10.0 Quads	25%	$109 Billion

*"At equivalence" means with all energy saving measures implemented whose cost is equal to or less than that of comparable new energy supply. The industries listed constitute a 20 percent sampling from which the figures for the overall industrial sector have been extrapolated.

Source: George N. Hatsopoulos, Thomas F. Widmer, Elias P. Gyftopoulous, and Roger Sant, *Internal Report* (Waltham, Mass.: Thermo Electron Corporation, April 1977).

percent of the opportunities are exploited, it is far more likely that the capital invested in energy conservation will turn out to be an excellent investment according to free market practices. How can we stimulate conservation in a most effective manner?

In addition to general inertia, one impediment to conservation is overregulation. Cogeneration of electricity seems to be one of the straightforward and effective ways to save energy. Why do we in the United States exploit this obvious possibility to such a small extent—in only 5 percent of all electricity production? Government regulations work very differently for industry generally than for utilities; the former are regulated only modestly, while the latter, including electric generating plants, are regulated much more stringently. Utilities and general industry seem to exist in different legal worlds. For the twain to meet would require organizational, legal, and technical expertise. It is, in a word, impossible.

Yet it must be done. In order to bring about the change, new regulations are being planned. The reasonableness of this approach might be illustrated by a medical analogy. Assume that a physician prescribes pills to help fight a bacterial infection, and that nasty side effects develop. The doctor then prescribes pills to combat the undesired side effects of the earlier pills. Shall this process continue indefinitely? And shall we look for the cure that is to cure the cure, and then for the cure that cures the cure of the cure? In medical practice this sounds outrageous. In regulatory practice it is considered progressive and enlightened.

The concept of energy conservation is simple and popular. Its execution by private citizens, industry, and government is complex and confusing. The following medicines might be prescribed:

☐ Rx for the Citizen: Understand one new fact each day and act upon it.
☐ Rx for Industry: Make one ingenious invention each year and follow through with its development.
☐ Rx for Government: Exercise moderation each second, and consider each night the consequences of your actions.

The difficulties of energy conservation seem almost insurmountable. But they can be mastered with time, with pain, and with reason.

PETROLEUM AND ECONOMIC PUZZLES

In which we learn that there is not enough oil, although new oil deposits are found when one looks for them. We also discover that if we try to reduce the price of oil and gas the price will increase. It is clearly demonstrated that the author does not understand economics—perhaps, in this particular case, with good reason. But apart from economic worries caused by manmade laws, the outlook for more petroleum need not be without hope.

It is generally believed that the world is running out of petroleum. Specifically, it is expected that America's oil reserves will be exhausted in a few decades. Oil and gas that have accumulated through geologic ages are now dwindling within the lifespan of a single individual. That fact and, even more significant, the way in which that fact is perceived, is shaping the climate of public opinion. And public opinion has generated some strong statements:

The Arabs are to blame!
The multinational oil companies are to blame!
The rapacity of modern industrial society is to blame!

None of these accusations is completely unfounded. None of them helps to solve our energy problems. But it may be well to consider them.

At the Persian Gulf it costs 15¢ to produce a barrel of oil. The sale price of that barrel of oil is close to $15, a high price made possible by the OPEC cartel. Cartels are illegal in America, but America's dedication to free trade principles does not prevent the formation of cartels in other countries. OPEC's existence may cause us to reexamine trade principles and possibly to accept regulations for the protection of the American economy, or that of other countries, from the consequences of OPEC prices. Later in this chapter we shall return to this question of regulations.

Big oil companies developed efficient methods of oil exploration and production. For a time this was beneficial to all, or at least it seemed to be. These oil companies were hit hard by the actions of OPEC. Their equipment was taken over in forced sales, in some cases for the value of only one week's production. The losses were eased because the oil companies carried equipment on their books at low, "conservative" values, having written off most of the investment value in order to reduce paper profits and taxes.

To me it seems appropriate to remember the contribution of oil to industrialization, which helped overcome poverty in many parts of the world, including America. If oil later is used up, this circumstance too will have served in establishing a new way of life, but we must find appropriate substitutes quickly.

We may ask how much time is left? How much oil remains to tide us over?

We must compare two time intervals in order to understand the situation. For how long a period in the future do we *know* there is oil enough for our needs? In how short a time can we increase production to cover the deficiency if we cease to import? The trouble is that the long time is not long enough and the short time is not short enough. It seems that by the time we can produce enough, the known reserves will be almost used up. This predicament suggests that we really are running out of oil.

Such a grim picture may, however, not describe reality. Actual oil deposits are far greater than known reserves. We need more exploration to find more deposits. But financial incentives are weak for finding oil to be used fifteen years hence, particularly since interest rates are high and the future of regulatory legislation uncertain. Meanwhile, even in regions where oil already has been found, it takes years to drill the needed wells. On dry land it takes five years to get substantially increased production; under the shallow waters off the coasts it takes ten years.

The delay may be even longer in Alaska, considering the difficulties of laying a pipeline over stretches of wet permafrost where structures are apt

FIGURE 6-1 *Because of Sierra Club objections, segments of the Alaska Pipeline were constructed to facilitate the migration of caribou. It has been observed, however, that caribou cross the pipeline everywhere, squeezing under, jumping over, and one baby even walking on it. (George Bing.)*

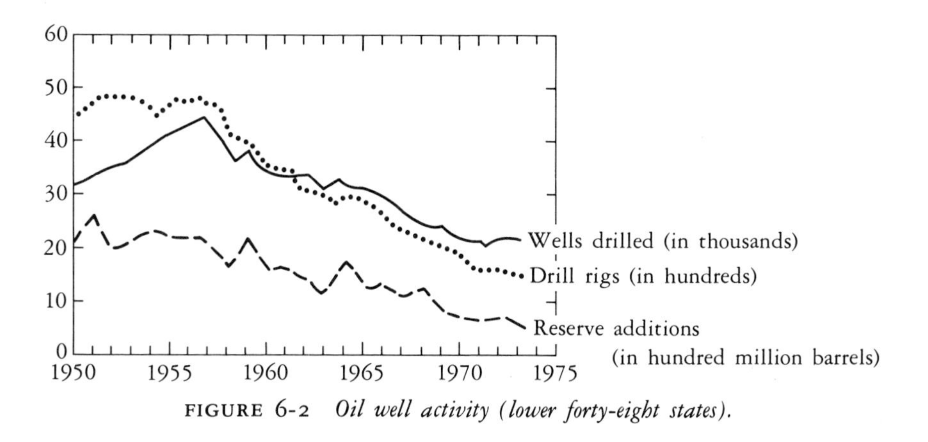

FIGURE 6-2 *Oil well activity (lower forty-eight states).*

to sink, caribou may stumble over the pipeline, and the Sierra Club may stumble over the caribou.

Increased activity by the oil companies could extend the period of assured oil supply. It could also shorten, to some extent, the time in which enough oil could be made available. Since this conclusion is contrary to conventional wisdom, it is necessary to look into the history of oil exploration.

In Figure 6-2, the amount of new oil found each year between 1950 and 1973 in the lower forty-eight states is shown. There has been a sharp decrease, with similar decreases in the number of wells drilled and the number of drill rigs in use, since 1956. It is important to note that in 1956 there were 5,000 oil rigs, but by the time of the oil embargo only 1,400 were operating. The three curves (oil found in new wells, number of wells drilled, and number of oil rigs used) obviously show the same trend. We seem to be suffering not from a scarcity of oil, but from a scarcity of oil rigs. (Incidentally, in recent years, the number of operating rigs increased to 1,700. As the oil-depletion allowance was cancelled and incentives for oil exploration decreased, the number dropped to 1,600.)

What caused the decline in the number of oil rigs in the 1960s? Oil was more easily found in the Gulf of Persia, so the rigs were taken abroad and were kept there. Old rigs were not replaced. The shortage is real and understandable.

One would think new rigs could be built, but this is not easy. A few years ago a price ceiling was placed on steel sold in the United States. As a consequence steel production for domestic use declined, but steel could be sold abroad at higher prices. Our steel companies made commitments to deliver oil rigs for work in the North Sea. A combination of economic and political circumstances reduced the steel industry's ability to deliver rigs and the oil industry's desire to buy them. It would certainly be in the interest of the United States to build these rigs. Yet United States government regulations impede what is desirable, though their harmful effects are unintended and unforseen. It follows: Think at least twice before you regulate.

The addition to reserves in each year shown in Figure 6-2 have been conservatively estimated. As a rule, the additional oil available is approximately three times greater than the original estimate. This addition has been taken into account in Figure 6-3 but not in Figure 6-2.

Another aspect of oil-exploration history is indicated in Figure 6-3, which shows the oil reserves added each year divided by the number of wells drilled in that same year. This figure is deceptive for two reasons. It includes the upward adjustments of reserves from older wells where, because the original estimate was conservative, much more oil is eventually produced. It also includes the conservatively estimated oil in newly drilled wells. In the last five years shown, the amount fluctuates about an average near 140,000 barrels. Since 1970, increased oil prices have stimulated more drilling; indeed, drilling activity has increased by 30 percent or more. At the same time, the increase in reserves divided by wells drilled in the same year declined to a low value of approximately 80,000 barrels per well, little more than half the previous figure for the reserves. This is due in part to the circumstance that reserves from new oil wells are un-

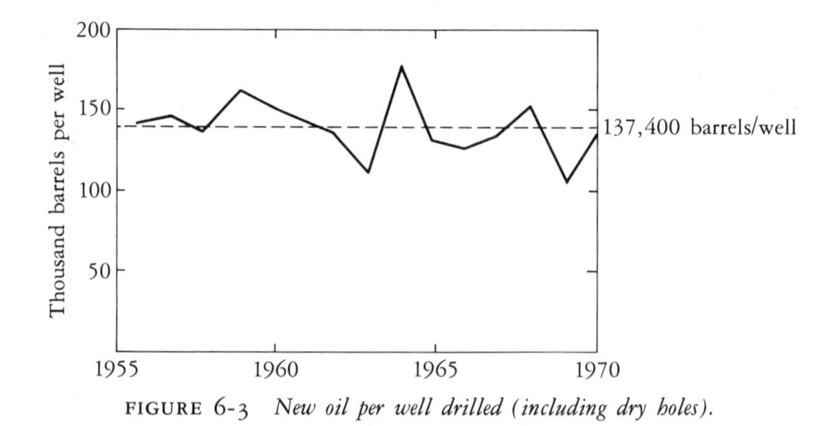

FIGURE 6-3 *New oil per well drilled (including dry holes).*

derestimated, as always has been the case. At the same time, increased reserves from adjustments of old wells are less important because fewer oil wells have been drilled in the recent past. There is also some real decrease in the oil found per well because, with increased price incentive, drilling is undertaken in locations where less oil is expected. Furthermore, wells are deeper than they used to be. In early days of oil exploration the average well depth was twenty meters; now it is two kilometers. The greater cost of deeper wells is offset, at least partly, by improved drilling methods.

The complexity of this discussion indicates how difficult it is to predict reserves and future production. Divergences of opinion and changes in estimates, even from the same agency, therefore become more understandable.

Now, having looked at the recent past, let us turn to the future. Where and by what technical means can we hope to find and produce more oil?

There is every reason to believe that oil will be found down to a depth of at least six kilometers. Beyond that depth, temperatures may be too high and porosities too low for oil to be found. To reach greater depths becomes increasingly expensive unless we continue to develop better drilling methods. One means would be to loosen rock with high-power water jets before cutting it with the drill bit. The Romans had a phrase for it: *Gutta cavat lapidem non vi sed saepe cadendo* ("Not by force is stone worn away, but by droplets that never cease falling"). But the Latin is inappropriate to describe the modern invention. We need not find the patience to wait until gentle droplets erode rock. The force, *vis,* of a water jet can be great enough to break the toughest rock in an instant. The drill rig need merely scoop up the debris. It seems likely that drilling in this manner can be both fast and cheap.

Drilling for oil on the continental shelf is another important technological development, as yet in its infancy. We have mentioned that with present imperfect methods it takes a decade to develop an oilfield on the sea bottom. This was clear in the case of oil in the North Sea. Traditionally, Texas towers, built on the sea bottom and supporting a solid platform above the waves, have been used. Replacing them with equipment installed on the seabed below wave action and operated by remote control could lead to the exploitation of vast additional reserves. This development will take time, and it may take several years to reduce costs.

So far we have considered oil under the continental shelf. As mentioned in Chapter 4, much more oil may be found beyond the steep continental slope on the gentler continental rise, approximately a mile deep. If this should be the case, the oil will certainly not be recovered by erecting conventional Texas towers. No skin diver will go to such depth; gases entering body tissues due to the pressure at great depth are destructive. Furthermore, to lower workers in a bathosphere would be much too expensive. The only feasible technology would be to place apparatus on the sea bottom and operate it by remote control. This type of technology may be useful not only for oil recovery but for recovery of other minerals, including manganese nodules. It may become an important part of the shape of things to come.

So long as we obtain oil by more or less conventional means, we are dealing with the near future. Exploitation of the continental rise is probably harder than large-scale utilization of solar energy, and may come later. It may, however, become an economic possibility that could keep the oil industry viable to the end of the twenty-first century.

With the oil spill in the Santa Barbara Channel, offshore drilling became highly unpopular. But we should remember the ubiquity of natural oil seepage. And the sinking of a single supertanker near the coast, with its enormous consequences, should be considered more seriously still. To import oil is no less hazardous than to drill for it.

We can choose a more positive approach. Oil spills can be prevented, and indeed though damaging, they are rare. Even more significant is the fact that if oil spills do occur they can be cleaned up. The easiest method is to prevent spreading by surrounding the spill with a corral built of pieces resembling logs. The oil can then be sipped up or otherwise cleaned up in a straightforward manner.

The corral will not work in heavy seas. There the best method may be to let appropriate bacteria consume the oil. Bacteria with a strange appetite for oil actually exist. They eat oil and nothing else. Oil came from living matter; these bacteria turn oil back into proteins of living cells. They occur naturally and live on occasional natural oil seepages. But their pasture is pretty lean, and few survive. When a juicy oil leak occurs suddenly, there are not enough bacteria around to consume it.

It is possible to culture these bacteria, which, at appropriate temperature, will flourish and divide in a vigorous manner. Then when an oil leak occurs this culture can be dropped on the spot, and the oil will disappear in a relatively short time. These bacteria can also be used to clean the holds of oil tankers, thus eliminating any temptation to wash out the holds with sea water and willfully cause pollution. Work on this use of bacteria has progressed well in Israel.

Another possibility, in urgent need of further research, is utilization of the hydrocarbons left when an oil well has "run dry." Running dry does not mean that no hydrocarbons are left underground. When approximately two-thirds of the original hydrocarbons remain in place, in the United States no additional oil can be produced. What remains is much too viscous to be moved by conventional means. It can be recovered, however, by one of three methods.

First is the use of *surfactants,* substances that act as lubricants and allow oil to slide out through small holes in the porous substance where the deposit lies. Of course, the surfactant itself must be inexpensive and must make possible the recovery of a comparatively great amount of oil.

The second technique is described by its picturesque name, "Huff and Puff." Hot steam is pumped down, the oil is warmed, and its viscosity is consequently lowered. Then comes the Puff—oil is pumped out from the same well.

In a third method, "fire drive," two wells are drilled. Air is pumped down into one and a fire is lit. From the nearby second well, oil is pumped out. Heat from the underground burning reduces viscosity, making the oil flow more easily. Another and perhaps greater advantage derives from the carbon dioxide produced by this underground combustion. The carbon dioxide dissolves in the oil, reducing its viscosity even further and allowing it to move to the second well. From there the oil is pumped to the surface.

Fire drive is an example of a future technology that I call underground engineering. The general idea is to move what used to be immovable. It may be possible to improve upon the fire drive in a number of ways. We shall discuss this technology in the next chapter where we consider a variety of fossil deposits.

In summary, one can say that with appropriate economic incentive and government action, oil obtained within the United States will remain a strong component of our energy economy for the rest of the twentieth century. Whether we can count on substantial amounts of oil beyond the year 2000 depends on research and on luck. Neither of these will be sufficient without the other.

What has been said about the United States holds generally for the rest of the world. The main difference is that the lower forty-eight states have been explored to a greater extent than most other regions. It would not be surprising to find massive deposits elsewhere. The great storehouse of the Persian Gulf may not be unique.

In the case of Israel, it has been remarked that Moses managed to find the only spot in the Middle East where there is no oil. Perhaps this statement is premature. Israel has not been explored to a depth of six kilometers, and much remains to be learned about its geology. Let's assume that oil should be found there at great depths and in great quantities. This would be good news and bad news.

Such an oil find would be good news in itself. The bad news would be that Israel would have to decide whether to join OPEC. I think this problem could be resolved. But then OPEC would have to decide whether to accept Israel, and with this consideration we are turning again toward good news. A further provocative question would be whether OPEC could ask Israel to join without recognizing that it exists. I think the good news outweighs the bad. So Israel should by all means look for oil. If oil were to be found, Moses would be splendidly justified. The possibility should also stimulate refinement of seismic methods of seeking petroleum at great depths.

In general, seismic methods have been limited to locating structures such as salt domes and other stratigraphic traps by monitoring the manner in which geologic layers reflect sound waves. The presence of the structure does not assure the presence of oil inasmuch as the trap is often empty. The reflection of seismic waves indicates only the *opportunity* to find oil.

Procedures have been developed recently that may identify the presence of gas if not oil, and the two are often associated. Both gas and oil are found in the highest portions of permeable porous rocks. If there is gas in the pores, it occupies the space above the oil. The presence of gas causes a slowing down of the propagation of sound waves, since gas is much more compressible. Careful evaluation of the reflected sound waves might give evidence of slower propagation of sound, and of the presence of gas.

Early in the century natural gas, produced incidentally with oil, was flared, that is, it was burned uselessly at the site. This practice still prevails near the Persian Gulf.

The Federal Energy Commission (FEC) took constructive steps toward utilizing gas in the United States in the first half of this century. An infrastructure of gas pipelines was established and gas was widely introduced into interstate commerce. It was cheap, it was clean, and it caught on. America started cooking with gas.

Unfortunately, FEC regulations outlived their usefulness. As late as the beginning of 1974 the ceiling price of gas in interstate commerce remained below twenty cents per million British thermal units. Since a barrel of oil represents six million Btu, this meant that the price of gas corresponded to little more than a dollar a barrel. We had become used to clean, inexpensive gas. Its low price stimulated demand for gas, but now the supply rapidly nears exhaustion. More gas could be produced, but at a higher price and with delays from causes such as the need for further exploration and the laying of new pipelines.

Yet natural gas has turned out to be habit-forming. Its use creates practically no smog. In a place such as Los Angeles, where air remains stationary over the city for long periods, people have to put up with the pollution they generate. Gas is vitally needed if Los Angeles residents are not to suffocate. But as present trends continue, clean gas becomes increasingly unavailable.

Can we bring in gas from other sources? It can be done by pipelines. Importing gas from Canada and Mexico via pipelines seems promising. Also, gas will be produced together with oil on the north slope of Alaska. An enormously long pipeline from the far-north region near Point Barrow to the lower states might still make some sense. It would seem easier to limit the pipeline to Alaska, and ship the gas from Anchorage to California. There are similar proposals to bring gas from Indonesia, where it is available and at present unused.

Unfortunately, the shipping of gas is expensive and dangerous. The gas must first be liquified. This necessity plus the transportation cost drives up the price to $4 per million Btu, almost twenty times the recent ceiling price.

Liquified natural gas, furthermore, is dangerous. It does not burn in the same manner as oil. The burning of an oil tanker affects the immediate area and may cause widespread, disturbing pollution. Liquified natural gas (LNG) may vaporize, mix with air, and then ignite several miles from its point of origin. Big fires, conflagrations, and even firestorms, in which oxygen is exhausted and people suffocate, cannot be ruled out. Ignition of gas from a small leak can cause a steady flame as the gas streams out and mixes with air. The flame then heats and weakens the orifice. Meanwhile, the temperature of the liquified natural gas within the container is raised beyond the boiling point. Pressure increases until a small explosion en-

FIGURE 6-4 *Rescue workers dig through ruins after a giant firestorm caused by two million gallons of escaped liquified gas in Cleveland, October 20, 1944. Spreading fire drove families from their homes as it gutted 29 acres, killing 133 persons, injuring 300, and bringing to a halt the early use of LNG. With modern usage this disaster, which was caused by use of metals not adapted to the low temperatures of liquified gas, will not be repeated. The much bigger amounts carried in tankers and stored in individual units makes careful study of LNG hazards mandatory. (United Press International Photo.)*

larges the orifice. Now the flame becomes brighter and more powerful. The process repeats itself. In the end the whole container could be destroyed.

What has been described is perhaps an improbable chain of events. (In the case of nuclear reactors, we are taking precautions against equally or even more complicated but less likely hazards.) Another conceivable accident which may be more probable should be mentioned, however. Ships do collide. In a collision the liquified natural gas may spill. If, as is likely in a collision, the gas catches fire, both ships, with all hands, are apt to perish. But this may turn out to be the least of the potential disasters.

Great precautions are being taken against collisions near a harbor. But a collision even several miles from the shore may result in grave consequences. When liquified natural gas pours out, it will float on the water. At the point of contact the liquified gas is heated by the water. This is apt to occur in an irregular fashion. Where heating is greater more gas is developed, and the pressure at the interface produces a wave which, in turn, increases the heat exchange. This self-reinforcing process results in rapid evaporation of the gas with consequent fast energy releases that we would call small explosions. Similar phenomena occur in the contact between liquid lava and sea water. When this process has run its course, the natural gas, no longer a liquid, will still have high density due to its low temperature. This layer of cold heavy gas can be blown shoreward by the wind. If the wind is strong, the gas will be dispersed and no harm will result. But if the wind is gentle, a mixture of gas and air will reach the shoreline. Once on land the mixture is apt to encounter a flame, start a fire, a conflagration, or possibly an explosion.

It is known that a mixture of natural gas and air is not flammable unless the volume of the gas is greater than 5 percent and less than 15 percent. In the former case there is not enough gas, and in the latter, not enough oxygen. Even so, parts of the cloud may well maintain this mixture at a great distance from the collison site. The possibility of a firestorm, killing not only by heat but also by exhausting the oxygen supply, cannot be ruled out.

It has been correctly argued that in confined places natural gas mixtures with air result in detonation. In fact, detonation will occur only if

FIGURE 6-5 *LNG Terminal at Cove Point, Maryland, receiving its first shipment of liquified natural gas from abroad, March 1978. From the carrier, the El Paso Sonatrach, at the dock, the LNG is piped to four storage tanks seen in the background.*

FIGURE 6-6 *Storage facilities at the Cove Point, Maryland, LNG Terminal. For protection in the unlikely event of a leak, tanks are surrounded by earthen dikes, reinforced and faced with concrete inside and out. As seen in this and the preceding photograph, the terminal is isolated from towns or dwellings.*

there is opportunity for pressure to be built up. In open surroundings this is not likely. However, it might occur if the mixture of gas and air is so extensive that the pressure, which spreads with the velocity of sound, cannot be dispersed.

The existence of all these possibilities does not mean that we must refrain from using liquified natural gas. In the present shortage some regions of the United States seem to have little other choice. It does mean, however, that the dangers of liquified natural gas should be studied much more thoroughly, and that the greatest precautions should be taken. Its dangers may far exceed the danger from nuclear reactors—not so strong a statement as it may sound. As we shall see in Chapter 10, nuclear reactors are remarkably safe. Yet until 1976 research on the dangers of liquified natural gas had amounted to no more than one percent of such studies concerning nuclear reactors. Now research is accelerating and safety is improved.

At present the shipping of liquified natural gas worldwide and in the United States is increasing. Single shipments are delivered in quantities that in the extremely improbable case of detonation (requiring a big-scale spill, widespread mixing with air, and a chemical reaction that might grow into an explosion) would deliver the destructive energy of almost a million tons of TNT. No equivalent situation exists with respect to nuclear reactors.

Modern terminal and storage facilities for liquified natural gas, such as those shown in Figures 6-5 and 6-6, are constructed with considerable care, since even a single big accident is unacceptable. Extensive loss of life must, of course, be avoided. We should also consider that a single big accident could force LNG transport to be shut down, and if we continue to increase our reliance on this resource the consequences could be most disturbing. The situation is similar to that in the case of nuclear plants. Prior to careful analysis of possible hazards and their control, objections on the basis of safety are justified. After regulations based on careful analysis are introduced, the objections should not be overemphasized.

The use of liquified petroleum gas (LPG) is at present limited to smaller quantities. Liquified natural gas consists mostly of smaller molecules, of methane. Petroleum gas is easier to liquify. Its molecules are bigger, consisting of propane and some butane (three or four carbon atoms rather than one in methane). Unfortunately, when it mixes with air it can explode more easily. Massive use of liquified petroleum gas should be delayed until its dangers are thoroughly explored.

In the long run the gas problem calls for two parallel solutions. First, to remove price ceilings and pay more for gas is much more tolerable than to run out of gas, to idle industries, and to render useless all the gas equipment in millions of households.[1] A price increase is more acceptable, since distribution costs that contribute greatly to the price will remain unchanged. If gas produced domestically could be sold for the price the market determines—for $2 at the wellhead, for instance—plenty of it would be produced within a few years. At the same time, the price in actual use would increase by a smaller percentage.

The second remedy is to look for radically new procedures by which gas or its substitutes can be produced. One procedure is to utilize "tight" gas formations, fields in which there is plenty of gas but where the pores

---

[1]Provision for gradual deregulation of gas included in the energy bill enacted by Congress and signed by the President in late 1978 is a step in the right direction. Even though the law is formulated in a cumbersome manner that will slow and complicate its implementation, the result may well be to ease the gas shortage within the next few years.

in the rock are so small that gas will not seep out at an economically recoverable rate. The rock can be broken by various means, including the use of water under high pressure, high explosives, or even nuclear explosives. These possibilities will be discussed in the next chapter.

There is no question that the age of ample petroleum simply pumped out of the ground will end. We may now be seeing the beginning of that end. On the other hand, the useful oil and gas supplies may be prolonged for another half century, a full century, or even a little longer. In the history of technology, it is not unusual for one resource to be replaced by another. What is important is that the transition should be carried out smoothly, without producing unnecessary shock effects in economics.

Shock effects have already occurred. The oil embargo of 1973 is remembered by everyone who does not try very hard to forget it. The gas shortage of the hard winter of 1976–1977 is more memorable still.

Congressional actions have been motivated, at least in part, by a perception that oil companies are responsible for the difficulties. It is imagined that steps must be taken to put an end to the improper behavior of the companies. In this regard we shall discuss three points separately:

☐ To what extent should one blame the big oil companies?
☐ What was the effect of governmental actions?
☐ What would be the consequences of the harsh measures which some people propose?

There is no question that the most important oil companies are big. The so-called Seven Sisters[2] are more famous for their power than for their grace. In today's climate of opinion, to be big is to be unpopular. Even so, the size of these multinational corporations made possible the introduction of effective methods of oil exploration and production throughout the world.

The oil companies have made mistakes that in fact have hurt the United States as well as the companies themselves. Following the dictates of free trade, they looked for oil where oil was most available, and invested in the areas where oil production was most profitable. The most important of these areas was the Persian Gulf. A situation developed where the Seven Sisters and other oil companies faced a dubious political future resulting in

[2]Exxon, Shell, British Petroleum, Gulf, Chevron, Texaco, and Mobil.

a dubious economic future. Even a big company has to struggle with the question of where to raise capital for continuing its business. In this case, capital is badly needed for exploration and production of oil and gas in the United States and in other regions of the world outside the sphere of the communists and the OPEC cartel.

Precisely at the time when initiative on the part of the oil companies would have been most important, initiative was inhibited by worries of a rapidly changing, clearly unfriendly political climate. Considering this fact, we have to return to the question of the effect of actions the government has taken.

Government actions were taken to relieve temporary difficulties. They accomplished this goal by introducing greater difficulties for the future. Price ceilings, for instance, shielded the American public from the harshest effects of foreign price increases. In the longer run these ceilings caused shortages, and they are apt to lead to even steeper price increases in future. By discouraging oil exploration in the United States, American price ceilings tended to confirm the virtual monopoly of OPEC.

But even more measures are apt to follow. A popular word is "divestiture." What is big must be cut back; let the oil companies divest themselves of other energy enterprises such as coal. In other words, if in trouble, reorganize. It is more reasonable to say that more changes should be avoided in the midst of troubles.

To introduce further harsh regulatory measures in this situation is a clear invitation to disaster. Free enterprise, coupled with relatively gentle regulations, has certainly worked as well in the United States as human institutions can be expected to work. Whether totalitarian systems can accomplish as much seems doubtful. But to introduce even more regulations in our society, which maintains the structure of a free enterprise system, appears to be an adoption of the worst features of the two opposite economic systems.

Let me return to the strongest statement repeated at the beginning of this chapter, "The rapacity of modern industrial society is to be blamed for the energy shortage." The fact is that American industrial society has, on the whole, not produced disastrous results for the average citizen. Furthermore, our industry is responsive, because it has to be, to public demands.

I cannot conclude with any firm statement concerning a reasonable oil policy. But I can raise a question. Was the whole trend of our regulatory procedures perhaps misdirected?

It is possible that in a closely interrelated world, where cartellike organizations exist (in the oil-producing countries, in Japan, and certainly in the centrally directed Soviet Union) a free economy cannot compete if it is hamstrung by rigid antitrust provisions. When oil companies fear to enter into cooperative ventures on tertiary recovery—that is, the recovery of vast amounts of hydrocarbons left underground—we may have driven regulations too far.

It may be feasible to return to a more free-wheeling economic order and resort to government interference only in the most obvious cases, for instance where the size of a needed activity is clearly beyond the reach of any private undertaking.

The alternative is to continue to introduce regulations, and when they fail to work, introduce more regulations to cancel the harmful effects of the earlier ones. We are keeping price controls to cool the economy, and then we apply tax rebates and other incentives to heat it up. This reminds me of a situation I found in one of our progressive enterprises. The building I visited had a heating plant and air-conditioning equipment. Both worked simultaneously to keep the rooms in the perfect balance of an ideal temperature.

I believe that in this case as well as in our policies on oil and gas, simpler solutions would be preferable.

It seems remarkable that in considering the urgent problems of these last two chapters, energy conservation and petroleum production, we are driven to the same conclusion: let's act, but let's put more confidence in our system of (almost) free enterprise.

MINING COAL AND OTHER FOSSIL FUELS

In which we find that there is plenty of the old fuel left in known deposits, and there is probably much more yet to be found. But much of this energy source is hard to obtain and difficult to use without damage to those who obtain or use it. These difficulties may be just as great as those associated with attempts to get energy by some radically new approach.

There is enough oil for decades to come. There is enough coal for centuries. This is true for the United States; it is also true for the world.

Coal fueled the Industrial Revolution. Richard Llewellyn's *How Green Was My Valley* reminds us that coal was not the cleanest nor the most pleasant foundation on which to build. Accidents in coal mines have taken a steady and terrible toll in human lives. Accidents occur still, even after great strides in safety measures. But normal mining conditions may have contributed even more to human suffering. Black lung disease is a terrible, often fatal, chronic condition caused by coal dust inhaled into the miners' lungs. Erosion and real environmental damage resulted when coal was mined on the surface without adequate remedial measures.

Yet coal was king. During its reign industrialization, with its successes and its injustices, obtained a foothold on our globe. The change that occurred can never be undone.

Now King Coal has lost his throne to the Sheiks of Araby. Is this a change for the better? To what extent can we change back to coal? Can it be done?[1]

First we must understand why coal has yielded first place to oil. In 1950 coal accounted for almost 60 percent of the world's energy production; in 1974 it accounted for less than 30 percent. In that quarter of a century, energy from coal actually increased 1.6 times. But the increase in total energy was more than threefold. In the same period, United States coal production actually dropped at first; then it increased, so that by 1974 it slightly exceeded the production of 1950. Total energy consumption almost doubled in the same time, however, and the percentage of energy from coal became hardly more than one-quarter.

Coal consumption dropped in western Europe, while energy consumption increased. In 1950 coal accounted for over 96 percent of the energy consumed there; at the end of the 24-year period it accounted for little more than 50 percent.

Germany was industrialized and became powerful using coal from the Ruhr, north of Cologne. But during the "industrial miracle" of West Germany after World War II, wages kept rising and coal became more expensive than oil. Some coal mines in the Ruhr were closed; pumps that

---

[1]See, for example, A. H. Pelovsky, ed., *Symposium on the Coal Dilemma: How and Where to Use It* (Denver: American Chemical Society and Cameron Engineering, Inc., 1977).

FIGURE 7-1 *(George Bing.)*

had kept them dry were stopped. Now those mines are flooded, the equipment rusted. It would be expensive to reopen them.

Chicago used to burn Illinois coal, which contained 3 percent, 4 percent, sometimes even 6 percent sulfur. Wind in the windy city did not blow fast enough to blow the sulfur fumes away, yet law required that Illinois coal be burned. Now a new law requires that the sulfur content be lowered to 0.5 percent. One extreme regulation has been replaced by the opposite extreme.

There is an abundance of coal in the eastern half of the United States, but much of it has a high sulfur content. Low-sulfur coal is plentiful in the West, particularly in Wyoming and Montana. But there are two obstacles to use of this coal: prejudice against strip-mining, and distance. Colorado and Utah are equally rich in coal. There the coal is at greater depth (the layers tend to dip farther down as we proceed toward the South) and there is less opportunity for surface mining.

The first obstacle and the one that should be least difficult to remove is the resistance to surface mining. There are many appropriate places in the western United States where we typically find a twenty-meter-thick coal seam with an overburden of another twenty meters. Most known deposits are not in steep hillsides. We can and must establish reasonable procedures for strip-mining of these deposits.

The necessary steps are straightforward. The overburden is removed and heaped up at the side. The coal is then scooped out. There are no underground miners involved, and no underground accidents. There is no black lung disease. After the coal has been hauled away the overburden is replaced.

An objection has been raised that in the dry climate of coal-rich Montana and Wyoming vegetation that is uprooted will not grow again; thus mining sites will be transformed permanently into badlands. There is a simple economic answer to this objection. The value of the coal mined on an acre is approximately half a million dollars. It would be entirely reasonable to demand that $5,000 of this revenue be put into rehabilitation of the land. The cost of two or three acre feet of water could be easily included, representing ample rainfall for a year, and in these regions, for even two years. There is no question that we could start the growth of some well-chosen vegetation. It may even be desirable to select vegetation other than the sagebrush that probably grew there originally.

This is not theory only. It has been put into practice by American Metal Climax Coal Company (AMAX), with headquarters in Indianapolis, for instance, at Belaire Mine in Campbell County, Wyoming. It is necessary, however, to get clear-cut agreement as to what is required and how it is to be enforced. Legislation passed in August 1977 appears to meet justified objections to strip-mining, and opens the road toward healthier development.

Transporting coal from the mine to the user is a more difficult problem. In mountain states the railroad may be the least expensive means. While railroad tracks are generally in reasonably good repair, the necessary gondola cars are not available. Vigorous expansion of mining operations, with proper surface mining equipment and proper transportation, will take five years or more. The cost is relatively moderate. An additional 600 million tons per year, which should double our coal production, will probably require a capital investment of only $30 billion. The resulting coal production would save the same amount in oil imports in one year.

Surprisingly, Alaska may be a good source of coal for the near future. On Cooke Inlet, a short distance from Anchorage, a considerable amount of coal is found right at the seashore, near the surface. The sulfur content of coal from this Beluga Field is not more than 0.2 percent, lower than the next best coal in the United States. This particular mine could be operating efficiently in as little as three years, if only enough customers could be found. It is remarkable that this valuable resource is idle in the middle of an energy shortage. It could make a considerable contribution around the Pacific Basin, in the western United States, Hawaii, and Japan.

When coal is near the seashore, as at Cooke Inlet, transportation could be much less expensive, but may involve novel procedures.

There is a glut of oil tankers. They could carry the coal, particularly if coal could be handled in liquid form. This can be done by suspending coal dust in water, from which it later has to be separated. It is also feasible to suspend coal dust in oil or in artificially produced methanol,[2] in which cases the whole suspension could be used as a fuel. A little more than half the heating value would come from the coal dust; more than half of the dollar value would be in the liquid.

Greater reliance on coal does not necessarily mean that we need more eastern coal mined by more coal miners. But coal mining in the East has the advantage that the mine is near consumers. This factor may mean less cost for shipping coal. It also may mean that electricity could be generated at the mine mouth, and then transmitted over shorter distances. All these practices would still require improved mining methods, greater safety for miners, more automation, and more modern equipment. In other words, we would continue an old development with its natural advantages and disadvantages. We have experience on which to build, but we should not expect any great progress except at considerable effort.

To get coal out of the earth and to deliver its energy to the consumer is only half the problem. How do we burn coal without serious damage to the environment? Can we convert coal into gas or oil? These questions are just as important as those about mining and transportation.

There are several parts to pollution from coal: particulate matter (smoke and ashes), oxides of sulfur, and oxides of nitrogen.[3]

---

[2]It has been found that coal particles tend to disintegrate in methanol (wood alcohol). Subsequently, dirt associated with the coal can be removed by centrifuging. This process eases handling and reduces the amount that must be shipped.

[3]To this list we should add a minuscule radioactive contamination that really does no damage. The remarkable fact is, however, that a coal-burning electric generating station is apt to emit more radioactivity than a corresponding nuclear plant. Opponents of nuclear power object to the latter (and lesser) form of emission, overlooking the former, which though greater is still insignificant.

FIGURE 7-2 *(a) The Sangre de Cristo mountains photographed in 1960, as seen from Los Alamos, New Mexico, about 200 miles downwind from the Four Corners coal-burning plant, which was put into operation after this photograph was taken. (b) The Sangre de Cristo mountains photographed in 1968, from the same vantage point. Most people in Los Alamos report that their view is typically veiled by haze since the Four Corners plant has been in operation. Only the small, generally micron-sized particles from coal emissions are carried over the 200-mile distance, the same particles that are apt to lodge in the lungs. (Courtesy of Wm. H. Regan, Los Alamos Scientific Laboratory.)*

We have a way to deal with smoke and soot. Particles in the smoke can be precipitated by charging them with electricity, then extracting them with an electric field. The apparatus is called an electric precipitator. It is inexpensive and gets rid of more than 99 percent of the particulate matter. So the problem is solved.

Or is it?

It is true that 99 percent of the *mass* is removed. What remain are particles the size of a micron, hardly bigger than the wavelength of light. The smoke has vanished from sight. We are not bothered by soot that is easily perceived by the eye. But the health hazard remains. The small particles that escape the electric precipitator can get into our lungs and lodge there. The improvement was cosmetic rather than real.

Research is underway on better precipitation methods. One approach that I want to mention is somewhat surprising. The reason that not all the small particles are precipitated by the apparatus is that electric charges will not stick to some of them. A particle to which sulfuric acid (H_2SO_4) has been added, however, can accept a charge, and can then be precipitated promptly.[4] The trick is to avoid too much sulfuric acid, using just enough to precipitate the particles. Any acid that is left over would be itself a dangerous pollutant.

The principle is the same as in a tale of the famous adventurer Baron von Münchhausen. He went lion hunting and had a good shot at a lion. Unfortunately, his powder was wet and, unable to fire, he ran away, incredible as it may seem for a man of his reknowned bravery, with the lion in hot pursuit. The lion was gaining on him. Münchhausen reached a stream and was about to jump into it when he spied a crocodile, its jaws

---

[4] As we shall discuss below, sulfuric acid is obtained from the sulfur dioxide emission from coal stacks at a later stage by adding oxygen and water.

(*a*)

(*b*)

agape, waiting for him. He looked back to see the lion crouched for the spring. Münchhausen ducked, the lion jumped over him, sailing into the throat of the crocodile, who bit off the lion's head and choked to death on it.

It is quite clear that two evils may cancel. But care must be exercised so that they cancel accurately.

The other contaminant (precisely the one we used in proper form as an antidote) now considered most dangerous is the sulfur contained in coal. During combustion sulfur is oxidized and becomes sulfur dioxide (SO_2), which has a bad odor but as far as we know is not especially dangerous. The trouble is that in the course of time sulfur dioxide picks up one more oxygen, and the resultant sulfur trioxide is harmful. It is most harmful if it manages to attach itself to the little micron-sized particles that can lodge in the lungs. In this way, corrosive sulfur trioxide (usually combined with water to form sulfuric acid or some metal oxide to form a sulfate) remains in contact with delicate tissues that it continues to irritate.

Pollution from coal fires was particularly noxious to London in its infamous pea soup fogs—they might more appropriately be called "coal soup fogs." In the 1950s one of these fogs killed thousands of Londoners. Consequently, burning of coal was greatly restricted, with special care taken to decrease the number of particulates. The situation improved radically, and now London's killer fogs are things of the murky past. It is possible that eliminating all particulates, including the small and practically invisible ones, is more important than eliminating the sulfur oxides.

After having neglected environmental considerations for more than a century, the United States, and to some extent the whole industrialized world, is rushing into action and particularly into legislation without waiting for clarification or emphasizing research needed for clarification.

In the United States, legislation for the protection of the environment proceeded without very careful analysis of the dangers due to pollutants. In particular, we lack solid evidence whether and to what extent sulfur oxide emissions are dangerous in themselves even after *all* particulates, including the small ones, have been eliminated. On the one hand, one may argue that there is always dust in the air and that sulfur emissions will inevitably find particles on which to settle and be carried into the lungs. On the other hand, if all particulates are eliminated from the stack emissions, the sulfur oxides may at least be carried away from populated areas; in time they will be precipitated so the damage they do may be reduced to a considerable extent.

To get rid of sulfur emissions would, of course, be a real advantage. Apart from using coal of low sulfur content, what can be done about it?

Approximately half of the sulfur resides, as a rule, in dirt from which the coal can be separated quite easily. The other half is contained in the coal itself. Its removal is difficult, and is best coupled with plans to "gasify" or "liquify" coal, a procedure we will discuss later.

Sulfur can be removed during or after the burning of coal. During burning, sulfur is transformed into sulfur oxides from which acids can be formed. They combine with alkaline substances, of which the most abundant is lime. So the idea is to add limestone powder together with some water and extract sulfur products from the burned material, finally obtaining harmless calcium sulfate ($CaSO_4$). Calcium sulfate, though, is a massive by-product and eventually we have to dispose of it.

The sulfur-lime contact may even be established during the burning process itself. In that case, coal and a much smaller amount of limestone are pulverized and mixed in a container with a perforated bottom. Air is then blown through the mixture. The blast lifts the particles, but is so regulated that it will not blow them away. The system is usually described as a fluidized bed, and it does look and behave somewhat like fluid. The procedure has the advantage of intimate contact between the sulfur and the neutralizing lime. It also provides the opportunity for burning to be guided by regulating size of the particles, amount of powder, and airflow.

In addition to particulates and sulfur, a third contaminant produced in burning processes, particularly at high temperatures, is nitrogen oxide. Here something should be said about the various oxides of nitrogen. In themselves they seem harmless, or almost so. But in combination with other chemicals they produce carcinogens. This may happen even by combination with the molecules in our own bodies. But it undoubtedly happens to some extent when nitrogen oxides get together with unconsumed hydrocarbons.

One example, discovered a few years ago, is a chemical with a rather long name, dimethylnitrozoamin:

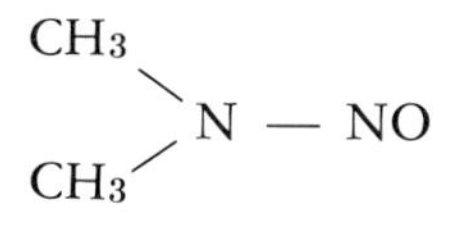

It forms at night but is decomposed by sunshine. In places where fuels are burned, there seems to be a correlation between the incidence of cancer and the formation of this substance.

The production of nitrogen oxide can be decreased by reducing flame temperature, and it might be practically eliminated by adding chemicals to the flame. It must be remembered that this contamination is less explored and more universal than the others we have mentioned. It is produced if air itself is exposed to a high temperature.

The radical cure for pollution by particulates and sulfur is, of course, to transform coal into something else: gas or oil.[5] Indeed, the gas used around the turn of the century was manufactured from coal. Natural gas came later.

We have described how coal, oil, and gas may have developed from decaying plants. The components of all three are essentially carbon and hydrogen. Some coal is almost pure carbon. Brown coal and lignite contain hydrogen together with carbon—as much as one or more hydrogens per carbon. In oil we have two hydrogens per carbon, and in gas the ratio is still higher, going up to four hydrogens per carbon in methane. The problem of transforming coal into oil or gas is that we will have to introduce more hydrogen. From this point of view it is better to start from brown coal, or even lignite, because they already contain more hydrogen. (Traditionally, hard coal, containing virtually no hydrogen, has been considered more valuable because of the heat developed per pound. For metallurgical purposes, practically pure carbon is usually preferred.)

Of course, hydrogen is readily available in water. In general, the process is to burn coal in the presence of water. At high temperatures hydrogen is transferred from the oxygen to the carbon. The result is a mixture of the ultimate burned products: carbon dioxide, half-burned carbon (carbon monoxide that can still be used as a fuel), various hydrocarbon compounds in which carbon and hydrogen are combined, and the remnant of air, which is nitrogen. It is easy to remove the carbon dioxide from this gaseous mixture before it is used. Nitrogen is more difficult to remove, and for this reason coal is often burned in oxygen rather than air, oxygen and nitrogen having first been separated by distillation at a low temperature where air is turned into a liquid. One must remember, however, to use air sparingly. Otherwise, the combustion is completed and no fuel is left.

It should be noted that the gas or oil we ultimately obtain from coal has less energy than the coal originally had, but sulfur and other impurities are left behind. The petroleumlike product is clean, and in many cases easier to transport. Also, it can be used in automobiles.

---

[5]Even this cure will not eliminate nitrogen oxides.

But is this product cheap enough? The answer, at least for the present, is no. What is proposed now was actually developed by the Germans in World War II. This obsolete technology will produce oil for $30 a barrel, or gas for an equivalent price of $5 per million British thermal units. Today the regulated price of gas is approaching $2, and the free market price exceeds $2.

Can improved technology reduce the price? Improvements are on the horizon, but they are not simple. One idea has caught the fancy of many people. Today we first turn coal into gas, then convert the gas into oil. Why not, they ask, perform the two operations in a single step? This course seems logical; the path we now follow is something of a detour. In coal the hydrogen to carbon ratio is low; it is highest in gas. In oil this ratio has an intermediate value. So the way we liquify coal today is to add a lot of hydrogen, producing gas, and then to remove the excess, thus getting oil.

Unfortunately, the direct path from coal to oil is much too slow a reaction because the structure of the hydrocarbon molecule has to be thoroughly rearranged. Living organisms show that it is possible for such rearrangements to be speeded up. The trick is to use catalysts, substances which do not get thoroughly committed in the action, but function as chemical "matchmakers." They bring the parties together in an appropriate way. (Incidentally, in living organisms the relevant catalysts are called enzymes. Without them we could not live.)

Finding the appropriate catalysts is difficult, but progress is being made. To use these catalysts on an industrial scale and in an economic fashion is an achievement that lies in the future, quite possibly in the distant future.

One difficulty is that there are many different kinds of coal. In some kinds substances are found that act more or less like catalysts. Other kinds of coal contain substances that may "poison" the catalyst we want to add. They latch on to the catalyst and, in this manner or some other, prevent the matchmaking. It would be much more practical to have a method that would work on any kind of coal. This method exists: the use of high temperatures.

The main trouble with coal gasification (converting coal into gas) and coal liquefaction (converting coal into oil) is that the apparatus needed for conversion is too expensive. At the same time, the apparatus yields too little product. Capital expenditure for a given output thus becomes too high.

Using higher temperatures can dramatically increase the rate of chemical reactions. The apparatus will be even more expensive, but the cost per unit product will generally be much less. The technology of high temperatures has made great progress, so there is hope for future use of this method.

There is, however, no certainty. At high temperatures everything happens faster. But what happens predominately is different from what occurs at low temperatures. High-temperature processes have been explored less, and we cannot say with real confidence that coal liquefaction and gasification will work. More precisely, we cannot say that they will work economically.

There is still another approach to coal gasification which, in principle, looks much better. Let us transform the coal underground *in situ.*[6] If I may be a little free with the translation from Latin, let's do the work where the coal sits. An expensive man-made retort would be replaced by the earth itself.

Advantages of this approach are obvious: no capital expenditure for conversion apparatus; no cooling water needed to safeguard man-made retorts (water is scarce in the big western coal fields); no mining, no accidents underground, no black lung disease. Deep, thick coal layers, as found in Wyoming, Montana, Colorado, and Utah, cannot be mined efficiently with present methods. With gasification *in situ,* leaving the dirt underground, we would bring to the surface a product similar to clean natural gas.

Unfortunately, *in situ* coal gasification has been tried for many years and it has not worked. In the minds of many people this settles the matter.

Actually, most of the experimental work was performed in Russia. It never has been discontinued, and at present it seems to yield some promising results.[7]

The negative results came from the United States. At the time that work was underway, natural gas was available for seventeen cents per million Btu, the equivalent of $1 per barrel. *In situ* coal gasification could not compete with natural gas at that price. At the much higher prices that

---

[6]See, for example, *Proceedings of the Fourth Underground Coal Conversion Symposium* (U.S.N.T.I.S., 1978).

[7]In Russia such gas is being used to fuel power plants, and the Soviets are attempting to export the technology. The process was advocated originally by Lenin, for the best reason: to save the coal miner.

prevail at present, *in situ* gasification may succeed, provided underground engineering is properly developed.

In situ coal gasification was attempted in one particular way that seemed promising. Let us consider a horizontal vein of coal. We drill two wells into the vein. Into the first we pump oxygen and, if there is not enough water underground to provide all the hydrogen needed, we add water.[8] Burning the coal will produce heat to transfer hydrogen from the water into the hydrocarbon compound. Then, so the plan would go, we pump newly formed gas out through the second well.

This plan works fine in its initial phases. Then the trouble starts. The hot gas burns its way through to the second well, making a channel in the coal vein near its top. In fact, buoyancy drives the burning process as high up as it can find coal. Once the channel through the coal is formed, nothing more happens. One can continue to pump oxygen into the first hole, and the same oxygen will come out of the second hole. The bulk of the coal has been bypassed and remains underground. The difficulty may be avoided, at least to some extent, by breaking up the coal between the two wells.[9]

There are at least two additional ways in which coal gasification may proceed underground. One possibility is to use coal veins that are strongly tilted. Oxygen and possibly water are pumped in at the lower end of the vein. Then fire, which produces gasification, burns its way up along the vein. It is particularly good if the coal will crumble easily. During the burning reaction, new fissures may open and most of the coal in the vein tumbles into the burning zone where it can be utilized.

In western coal there is a conspicuous formation called the "Great Hogback." Its name is descriptive of its appearance. It was formed by layers of the earth being thrust upward so that what once was horizontal now approaches a vertical position. Coal has been mined in many parts of that formation. One of these mines at Harvey's Gap near Glenwood Springs, Colorado, caught fire seventy years ago. The fire is burning still. The air supply has not been found, but hot gases, apparently completely burned, do emerge. We understand far too little about the way this burning process proceeds. We do not even know precisely where it takes place.

---

[8]This plan has some similarity to the tertiary recovery of oil, called fire drive, described in the preceding chapter.

[9]It may seem peculiar that burning proceeds in the presence of water. Actually, water puts out fire if it is present in quantity sufficient to lower the temperature or exclude oxygen.

FIGURE 7-3 *The Dakota Hogback, Front Range, Central Colorado. The eroded face seen in this photograph is an emerging portion of a steeply tilted layer. (Courtesy of Jack Rathbone, Rocky Mountain Association of Geologists.)*

The process, of course, is exceedingly slow and in its present form is completely useless. But by studying it we might well learn more about chemical reactions that occur underground.

The second approach is, in a way, the opposite of the first.[10] One tries to locate a deep, thick underground coal deposit. It might be located two hundred meters or even a kilometer below the surface; its thickness may be fifteen meters or more. Here the intent is to start the fire at the top of the layer. Burning would then proceed downward as the coal at the top is consumed. Oxygen is fed to the top; gas is drawn out near the bottom. Coal is virtually impermeable, and in order for the process to work, the coal must first be converted to rubble. This could be done by repeated use of high explosives deep underground, in quantities of about one ton per shot, an amount still small enough so that effects on the surface will be all but unnoticeable.

---

[10]Gary H. Higgins, *A New Concept for* in situ *Coal Gasification, Revision 1* (Lawrence Livermore Laboratory Report No. UCRL-51217 rev. 1, 1972).

The potential difficulty in this process is that, as the fire spreads downward, heat advancing before the fire may soften the coal. Under the pressure of overlying layers the softened coal may then be pushed downward in such a way that the gaps will be filled in and the rubble will return to a state of solid coal, with no permeability.

It is not certain that either of the described methods will work in a really satisfactory fashion. It is quite possible that some other approaches to *in situ* gasification will be preferable. These questions cannot be settled without a considerable number of experiments carried out in the field. It is not clear what the optimal depth is, nor do we know what kind of coal may most easily be turned into gas.

As a general rule, gas obtained at the surface contains a great amount of carbon dioxide, which need not be removed if the gas is to be burned locally. In fact, in that case air might be used rather than oxygen in the underground burning. Heat generated near the coal field can be used to produce electricity, and electricity in turn can be made available at a distance of one hundred miles or more.

Often, however, it is more desirable to transmit gas for use at some distant location through a pipeline. This process becomes too expensive if the gas contains many inert ingredients such as carbon dioxide or nitrogen. What we get from the coal field is a dilute or low Btu product. It does take surface processing to change low Btu gas to high Btu gas. If we plan such a process it may be better to eliminate the use of nitrogen in the first place; that is, we should feed the fire with oxygen rather than with air. It means that we must remove the carbon dioxide, and it also means that the remaining gas should be transformed in man-made reactors so that gas transmitted through the pipeline will have a higher energy content per unit volume. Therefore, *in situ* coal gasification need not really mean that we get rid of all processes on the surface. It means only that these processes are simple, less expensive, and that they need not deal with some of the impurities that hopefully have been left underground.

The amount of effort that has gone into *in situ* coal gasification is remarkably small, even though this process could yield an abundant supply of clean fuel. There are many similar underground processes which should receive considerable attention.[11] All of them have in common that they exploit fossil fuels that otherwise would not be available.

---

[11]See, for example, *Energy and Mineral Recovery* (U.S. Department of Energy Document No. CONF-770440, 1977).

One important possibility is the use of gas in "tight" formations. The word "tight" means that gas is found in small cracks and capillaries, so it cannot be moved easily from its location in the rocky substrate. We have to split the rock and turn it into rubble. Splitting has been attempted by hydrofracturing—that is, by pumping water down under high pressure. The rock is apt to split along vertical planes. It works, but on the whole it doesn't work well enough.

The next possibility is to use high explosives. The result is again ineffectual.

The obvious answer would seem to be a big explosion, in fact a nuclear explosion. This solution has been tried three times and progress has been made. The technique could probably be perfected, resulting in a considerable additional supply of natural gas. In this case the main difficulty is psychological.

The second of the three underground nuclear shots was carried out in western Colorado near the town of Rulison. We explained the purpose of the experiment and the necessary safety measures to residents of the area. For complete safety, a circle of a few miles radius had to be temporarily evacuated, and the local people went along with the suggestion. But representatives of the antinuclear movement appeared on the scene. They looked like wild west characters, but came mostly from New York. Five of them sat down right on top of the drill hole through which the bomb had been lowered. They would not be evacuated.

I was present, and I offered to sit down with them. This was not as heroic as it sounds. The sit-down occurred on top of a mesa 8,000 feet above sea level; the nuclear explosion, occurring at sea level, would have been separated from us by more than a mile and a half of rock. To me it would have been a worthwhile experience to stay, but I was not allowed to do so.

Instead, a helicopter was sent, offering a lift to the protesters. Three chickened out. Two stayed, standing on the drill hole. They were advised, "At shot time, bend your knees." This had no religious significance. The purpose of bending the knees was to absorb the shock, which, although by no means destructive, would have been noticeable. The most sensitive part of the human body is the spine, and if a shock hits you from below while your legs are straight, you might get a backache. We never heard of the protesters again. One may hope that they bent their knees and did not get a backache.

Rulison succeeded in giving useful information, as did one more shot at Rio Blanco, Colorado. But by that time antinuclear sentiment had become too strong, and a proposition was passed in Colorado forbidding any further experimentation with nuclear explosives without explicit statewide consent.

So far we have discussed production of gas by *in situ* processes. Changing coal into liquid oil underground appears difficult. But existing oil can be extracted from the earth more efficiently than it is now. Possible ways to accomplish this were discussed in Chapter 6 in connection with tertiary recovery of oil from old wells.

Many other fossil substances underground can serve as fuels. The best known of these is oil shale, an intermediate state of decaying marine vegetation that might turn into oil in the course of additional millions of years. In oil shale, a hydrocarbon that is not yet oil is anchored to a very small limestone particle. So oil shale is really neither oil nor shale, and it contains little if any of these two substances. If oil shale is heated, its hydrocarbon chain, called *kerogen,* does turn into a liquid that is similar to oil.

Mining oil shale and then heating it in retorts has been proposed. Like goal gasification on the surface, this process would be very expensive. One would have to mine enormous amounts of oil shale, build, use and cool expensive machinery, and be left in the end with a great amount of rocky substance which, if left lying around, might disfigure the landscape.

In Israel oil shale is found south of the Dead sea, not far from Arad. The Israelis are planning a remarkable use of this oil shale. They want to burn it without processing it. Oil shale has in fact been called the "stone that burns."[12] They plan to use ashes from the burned oil shale as the main ingredient of cement.

The outcropping near Arad is part of a big basin under the Dead Sea and surrounding it to the east and west. It is suspected that there are considerable amounts of oil shale in this basin. Most of it is oil shale without the oil and is called "mottled stone." Much of modern Jerusalem is built of this good-looking mottled stone.

A detailed analysis of the stone, including an accurate analysis of its isotopes,[13] showed that this stone had been at a high temperature a few million years ago. This is remarkable because the entire basin is made of

---

[12]There is a legend in Colorado about an early settler who built a cabin and invited his friends to a housewarming. He lit the first fire in his newly built fireplace. To everyone's surprise the fireplace itself caught fire and the log cabin burned down with it. The fireplace had been built of oil shale.

[13]See footnote 12, p. 22.

FIGURE 7-4 *A section of the Green River Formation of oil shale that extends through Colorado, Utah, and Wyoming, photographed at Anvil Points, Colorado. There are several hundred feet of oil shale in the formation, which contains an average of thirty gallons of oil per ton. This solid formation probably represents one of the stations along the long road from marine vegetation to oil. The development into oil could be accelerated by modern technology. (Courtesy of Mobil Oil Corporation.)*

sedimentary rocks and these stones have not been in contact with hot lava or magma. It is likely that in the past the whole basin was made of oil shale, that it caught fire, and that most of the hydrocarbons burned. This explanation accounts for the evidence of high temperature. It also suggests that extensive underground burning took place. In that burning process the oily substance was not only driven out of the limestone substrate, but was actually consumed.

We have more oil shale in the United States than all the oil of Arabia. The best hope of utilizing it is to reproduce under artificial conditions what happened in the earth a few million years ago near the Dead Sea. It would be essential to limit the amount of air with which the oil shale is burned. The process is similar to the one described above for coal gasification. By the use of explosives we can turn oil shale deposits into rubble. Air is then introduced at the top of the rubble zone and the shale rubble set afire. As the fire proceeds downward, preliminary heating of the oil shale will free the oily substance, permitting it to collect at the bottom. The fire is sustained by nonvolatile residues that contain carbon and very little hydrogen. This process has actually been tried near Rifle, Colorado, a location on the southern rim of the Piceance Basin. That deserted region contains, in some places, high-grade oil shale in a thickness of five hundred meters or more.

There are extensive oil shale deposits in other parts of the world. Brazil is the outstanding example. In Estonia the material is used as a fuel without prior modification. Perhaps oil shale could be used in many places to alleviate the worldwide energy problem.

It is hardly possible to forsee every way in which fossil fuels can be utilized. Another rich resource is the tar sand found in Alberta, Canada, where sand is mixed in with high-viscosity hydrocarbon. It is difficult to obtain the hydrocarbon without the sand—a situation quite different from a viscous oil contained in a solid, practically immovable matrix.

In the eastern part of the North American continent there are many bituminous deposits that are in some ways similar to coal and in other ways to oil shale. It is possible that with the help of underground engineering we will develop ways to use the heating value of these substances.

Big quantities of methane seem to be widely distributed near the surface of our globe. This gas can be pumped out of coal deposits, thereby decreasing the danger of explosions when the coal is mined. The quantities of methane are big, but the rate of obtaining it is slow.

In some places methane is present in deposits of hot pressurized water, on the coast of the Gulf of Mexico, for instance. It is not clear to what extent this methane can be extracted economically. The matter is further discussed in Chapter 12.

The largest methane supplies have been reported in the Arctic at great depths, under pressure of hundreds of atmospheres.[14] This methane,

---

[14]One atmosphere is equivalent to one kilogram per square centimeter, or—in units that are hopefully obsolescent—14.7 pounds to the square inch.

combined with water molecules into a fairly regular structure called *clathrate,* is virtually a solid. A huge source of natural gas would become available if Arctic methane could be made accessible through wells. Unfortunately, no practical solution has been proposed for gaining access, and particularly for transforming the methane, under controlled conditions, into an easily flowing form.

At the beginning of Chapter 3 we mentioned that the source of natural gas may not be connected in all cases with the decay of living material. The fact that we find great amounts of natural gas in a variety of forms might turn out to be evidence pointing in the same direction. If this hope should be justified, and if we can develop improved methods of finding natural gas based on this different assumption concerning its origin, the shortage of this valuable substance may actually be of a temporary nature.

The coal and oil we have consumed so far represent only a fraction of what is still available, especially if we include sources that can be found ten, twenty, or thirty kilometers under the earth's surface. Finding a way to utilize this supply without excessive cost and with safety would be most valuable.

We should not conclude this discussion of fossil fuels without mentioning one more contaminant, carbon dioxide. This substance is nonpoisonous, harmless, and present in relatively small quantities; nevertheless, the total amount in the world's atmosphere is massive. Burning of fossil fuels and deforestation have increased the amount by 20 percent of the value it had two centuries ago. Incidentally, what matters is not the total fuel burned, but the rate of burning. Carbon dioxide remains in the atmosphere for décades or centuries. It is then dissolved in the ocean where it is precipitated on the bottom as $CaCo_3$, limestone, to join the geologic deposits.

The reader may wonder why we worry about carbon dioxide if it is harmless. It is a gas that freely admits sunlight, but tends to retain our planet's surface heat by preventing the escape of heat radiation into space. This "greenhouse effect" will be discussed in connection with solar energy in Chapter 12. More carbon dioxide may mean a warmer climate on the whole earth. This is not certain, however. Meteorology is a tricky business, and there are other effects which could decrease or even reverse the warming effects of the carbon dioxide blanket. Certainly more research on meteorology is needed in order to predict the end result if we burn more and more coal. There is possibly not enough oil in the world to make a real difference.

A warmer climate would have many effects. Polar icecaps may melt, partly because according to simple, perhaps much too simple considerations, there would be a greater increase in temperature in the polar regions. Near the equator great amounts of heat can be absorbed by evaporating water; the polar regions do not have a similar effective method to dispose of additional heat energy. Thus the poles may warm up more rapidly and icecaps may start to melt, probably at a slow rate. Near the South Pole, where great amounts of ice are in contact with water, the rate of melting might be enhanced. As a result the sea may rise as much as thirty meters, inundating New York, Tokyo, Leningrad, London, practically all of Florida, and much of Holland. One could not favor such submersive activities.

Some anticipate another difficulty. An unprecedented increase in the carbon dioxide content of the ocean may lead to a slight increase in acidity and prevent the formation of seashells. If that should be true, future consequences are difficult to estimate.

Before advocating abundant use of fossil fuels, we should understand the ultimate consequences of doing so. We are unsure of what potential effects may be; we do not know how fast they could occur. A thorough analysis, including a determined attempt to understand the causes of changes in climate, is badly needed.

That man affects his environment, and that we should take these environmental changes into account, is generally accepted today. What is often not realized is that we rush to conclusions and regulations concerning the economy even before we have thoroughly understood the eventual consequences of our actions. It is by no means clear which will happen first: the exhaustion of our fossil fuel reserves, or an overproduction of carbon dioxide due to the accelerated burning of these fuels.

Fortunately, we may find other sources of plentiful energy. One possibility that would preclude the production of carbon dioxide is to grow plants for fuel. Plants grow much more rapidly when carbon dioxide is added to their environment. We might use the exhaust from power plants that burn fossil fuel, or recently grown fuel, to feed carbon dioxide to plants. Thus the equilibrium of nature could be reestablished.

An even simpler procedure would be to set a limit—and not a particularly low limit[15]—on the burning of fossil fuels. Energy could be

---

[15] An increase in coal consumption might be particularly important to the developing nations.

obtained from nuclear fuels that produce by-products in small enough volume so that we can handle them. But nuclear energy is a new source not understood by the public as well as the old familiar fuels. Many objections have arisen against it. In the next few chapters these worries will be enumerated, and the subject illuminated by an account of the experience and expertise available in this controversial field.

PART III

NUCLEAR ENERGY

THE BEGINNINGS OF ATOMIC ENERGY

In which the reader will learn many facts that today may seem irrelevant, but that remain interesting because the history of atomic energy is the background for deciding whether and how to use the great forces of nature that have become accessible to us in the recent past.

I shall try to describe the earliest history of atomic energy. What was, in a way, its most remarkable chapter concluded in 1945. This is now a third of a century ago, and it is a little hard for me to realize that some of my readers had not even been born at that time. In my own eyes I now appear as something of a relic, if not a fossil; and I would like to be as useful as a fossil can be in giving information about the past.

The story starts in Canada, where Professor Ernest Rutherford taught at McGill University. He got into trouble because in interpreting radioactivity he specifically stated that elements would be transmuted by the emission of alpha particles.[1] A chemistry professor of that earlier generation denounced him as a fool, ignorant of an elementary fact of chemistry: that elements are immutable. Rutherford did not mend his ways. He continued his revolutionary pursuits in Manchester, England, where he made the basic studies that led to understanding the structure of the atom.

I never met Lord Rutherford, but I heard him speak a few years before World War II, in the mid-1930s. He seemed indignant and he sounded incensed. He mentioned the suggestion that great amounts of energy stored in the nucleus might be put to practical use, denouncing as visionaries all who would believe it. He asserted that these energies are safely contained in the innermost part of the arom, protected by the strong electric repulsion between positively charged nuclei.

I did not know then the reason for his indignation, but I happened to find out a few weeks later. Shortly before Rutherford spoke at that meeting, Leo Szilard, another man of Hungarian origin and one of my good friends, had gone to him and had tried to tell him how the energy of nuclei might be used. He did not get past a few sentences—Rutherford threw him out of his office.

This is how generations change. The crazy young genius who worked in Canada had turned into the omniscient lord of the hallowed Cavendish Laboratory.

What Szilard tried to tell Lord Rutherford has become history. He said that although positively charged nuclei have the greatest possible difficulty in approaching each other, one building block of these nuclei, which does not occur in nature in the free state, can approach any nucleus;

---

[1] Alpha particles are nuclei of helium atoms. They are very stable. Earlier we discussed their formation when the universe was half an hour old.

that is the neutron. The only trouble is that it is extremely difficult to make a neutron. When a neutron hits a nucleus, more energy is released than a million chemical reactions could produce. Unfortunately, a neutron is lost in the process and cannot be replaced except by a truly excessive expenditure of energy.

Szilard asked himself a question: "Might there not be a way in which the production of neutrons could become self-propagating, in which one neutron could make two and two could make four? A divergent chain reaction could start which then could involve a whole lump of matter." Pursuing these speculations, he concluded that for this process to occur you would need a sizable amount of the material in which the chain reaction was to take place; otherwise the neutrons would escape without being effective in self-propagation.

Szilard also became quite specific, and said that in order to try to make such material one should look at three elements: beryllium, thorium, and uranium. We now know that uranium is the starting point, thorium is probably the fuel of the future, and only in the case of beryllium was he wrong. He believed, rightly, in Einstein's equation $E = mc^2$. He knew that by measuring the mass of nuclei one could learn their energy content, and beryllium seemed to have so much energy that it could fall apart and emit a neutron. His logic was sound. But he relied on faulty measurements of the mass, which were generally accepted at that time, made in experiments performed by the great British physicist Francis W. Aston.

While Szilard—who, like myself, had escaped from Nazi Germany—was pursuing these ideas in London, a very different approach was being taken in Italy by a man whom I had met previously and with whom I was later to work for many years, Enrico Fermi. It was Fermi and the group of young physicists around him who used neutrons to bombard every element in the periodic table. They observed neutrons being attached to the elements to create new kinds of nuclei, and they observed the resulting transformation, *beta radioactivity.* This kind of radioactivity is a spontaneous change in a nucleus that has an excess of neutrons.

In this systematic work Fermi at last arrived at the heaviest element known at that time, uranium. When the uranium nucleus was bombarded with neutrons he saw something exceedingly peculiar. Instead of the formation of one, or two, or three radioactive substances, he observed scores of them. No one seemed to understand. It was natural to suspect that elements heavier than uranium had been made. Fermi hoped that he had

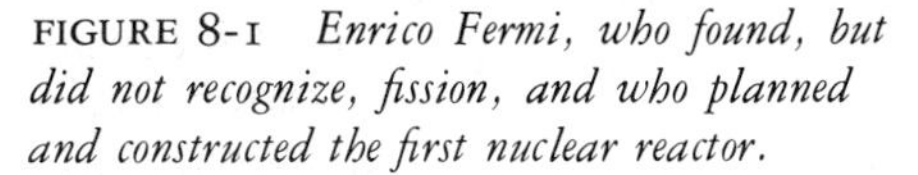

FIGURE 8-1 *Enrico Fermi, who found, but did not recognize, fission, and who planned and constructed the first nuclear reactor.*

opened the door to transuranic elements. For doing so (and for his many additional contributions) he received, a few years later, the Nobel prize, which he used to escape from Italy and come to the new world. The Nobel prize was well deserved by Fermi, but not for isolating transuranic elements, which were, in fact, not what he had observed.

The story is peculiar and in many ways ironic. Fermi announced his unexpected results. Soon afterwards he received a letter from a German chemist, Ida Noddack. Many years earlier she and her husband had, they believed, discovered two new elements, which they called masurium and rhenium. Of these two, rhenium in fact exists. They were mistaken about masurium. Indeed, masurium would have filled a place in the periodic table of elements that carry charges between one and ninety-two, but it does not exist in a stable form. It has since been artificially synthesized by one of Fermi's students, Segré, and its name, technetium, characterizes its artificial nature.

Mrs. Noddack's letter to Fermi said, "It is obvious what you are observing. You have given some extra energy to the uranium nucleus and because of the big charge that this uranium nucleus carries, it wants to split in two, to perform fission."[2] Actually, I do not think she used the

[2]This statement and those shown in quotation marks elsewhere in this chapter are not actual quotations, but fictional reconstructions based on the facts.

word "fission"—certainly not, because the letter must have been written in German. But the idea was there.

As an excellent theoretical physicist and as a freshly baked experimentalist, Fermi refused to believe her. He knew how to calculate whether or not uranium could break in two. When the uranium nucleus, which is more or less a sphere, begins to distort and becomes an ellipsoid in order to divide, the energy will increase. Only after separation is well underway does the electrostatic repulsion between the two halves of the nucleus begin to predominate and cause the two fragments to rush apart. In order for fission to occur the particles in the nucleus must "jump over a barrier."

Fermi knew that in the new theory of atomic physics a system can jump over barriers, but he also knew that this could happen only with small probability, the probability acutely dependent on the height and width of the barrier. He performed the calculation Mrs. Noddack suggested, and found that the probability was extraordinarily low. He concluded that Mrs. Noddack's suggestion could not possibly be correct. So he forgot about it. His theory was right, like that of Szilard, but also like Szilard's, it was based on the same wrong experimental information. Aston's experiment had at that time introduced a systematic error into calculating the mass and energy of nuclei.

Fermi had another shot at discovering nuclear fission, but he missed this second opportunity. Uranium emits alpha particles. He asked himself a pertinent question: "When I give uranium extra energy by bombarding it with neutrons, will I get alpha particles with excess energy, that could go farther before they are stopped?"

While looking for these extra alpha particles he did not want to be bothered by all the alpha particles that uranium emits spontaneously, so before bombarding uranium with neutrons, he covered his sample with foil just thick enough to stop the regular alpha particles but not thick enough to stop particles of any excess range. A conscientious man, he did that every time, which was unfortunate. Had he forgotten only once, he would have seen something startling. He would have seen the fission products coming apart with no longer range than a regular alpha particle, but with vastly more energy, giving far more ionization and bigger signals in the counter than anybody had ever seen. Had he seen that, I am sure he would have discovered fission. But he never forgot to put the foil over his sample.

A less noted physicist, Professor Paul Scherrer, in Zurich, tried the same experiment on thorium that Fermi performed on uranium, but he did not protect his sample with foil. He saw the big pulses of ionization caused by the fission products, and promptly called up his man in the electronics lab to ask, "Why is my chamber sparking today?"

According to all logic, fission should have been discovered about 1935. If it had been, it is highly probable that the Nazis in Germany would have taken hold of it, because they were interested in preparing for war. This eventuality might have resulted in the first atomic bombs being perfected in Germany, and could have given a completely different, and indeed catastrophic, turn to history.

As it happened, fission was discovered in the Kaiser Wilhelm Institute in Berlin in 1938, not by physicists but by two chemists, Otto Hahn and Fritz Strassman. They investigated the substances that Fermi, in his now famous mistake, still believed to be transuranic elements. One of these elements looked much like iodine, another much like barium. The more they tried to separate the iodinelike radioactivity from normal iodine or the bariumlike element from barium, the less they succeeded. Finally they were forced to conclude that the capture of neutrons by uranium produces the common lighter elements barium and iodine—or, more precisely, their isotopes. It was then no longer possible to avoid the conclusion that the uranium atom actually splits in two.[3]

The news was brought to Copenhagen by one of the refugees from Austria, Lise Meitner. She and one of her relatives, Otto Frisch, both very good physicists, looked for the fission products that Fermi never saw and Scherrer never recognized. They promptly found them. The news about fission spread like wildfire through the world of physicists.

My good friend Leo Szilard had already developed a theory, and it was known to all of us. *If* in this violent reaction of the splitting of the nucleus a sufficient number of neutrons got free, then nuclear energy, many million times more potent than chemical energy, could indeed be used, and used most simply in explosive devices. The question was, were additional neutrons produced during the fission process? An emphatic answer was found by Szilard within a few weeks. Neutrons can be slowed down by repeated collisions with nuclei of hydrogen atoms. When they are emitted from a nucleus they move at first with high velocity. Szilard slowed down the neutrons. When he bombarded uranium with these slow neutrons he

---

[3]The suggestion was actually made first by another German physicist, C. F. von Weizsäcker.

FIGURE 8-2 *Third Washington Conference on Theoretical Physics, February 18, 1937. In a similar conference two years later, Niels Bohr (front row, second from right) brought the news about fission, and the same day uranium fission was experimentally confirmed by Merle Tuve (back row, far left). Hans Bethe (front row, center) played a considerable role in the development of nuclear energy. Eugene Wigner (left, standing in front of Tuve) was the chief developer of nuclear reactors. The 1937 conference was devoted to problems of the properties and interaction of elementary particles and related questions of nuclear structure. In January 1939, all of these participants were again present. Together with Bohr, John Wheeler (third row, right, seen just above Bohr's head) prepared the first paper on fission. The conferences were sponsored jointly by Carnegie Institution and George Washington University. (Courtesy of John Archibald Wheeler.)*

found that new fast neutrons were produced. They must have originated in the fission process.

When he found the fission-produced neutrons, Szilard actually called me long distance from New York. I lived in Washington, D.C. I was playing the piano with a friend who was a violinist. We were in the middle of a Mozart violin sonata when the telephone interrupted. Szilard told me, "I found the neutrons."

That was in March 1939. To me, having talked with Szilard previously and with Fermi many times, it became very clear that this discovery meant the prospect of a great and fateful change.

I was present a few weeks later when some physicists got together

FIGURE 8-3 *Leo Szilard, whose initiative started the Manhattan Project and who never said anything that was expected of him.*

with Niels Bohr at Princeton. We tried to convince him that we should go ahead with fission research but we should not publish the results. We should keep the results secret, lest the Nazis learn of them and produce nuclear explosions first. Bohr insisted that we would never succeed in producing nuclear energy, and he also insisted that secrecy must never be introduced into physics. He was also the first to recognize, together with John Wheeler, that the really important substance for nuclear explosives is uranium 235.[4] It is, he said, extremely difficult to separate this isotope (same atom, different nucleus) from the much more abundant uranium 238. Atomic explosions may not be feasible, after all. In the end we convinced Niels Bohr that the difficulties could be overcome and the secret should be kept. We still did not manage to impose secrecy, however, because one man, Frédéric Joliot, in France, would not agree. Secrecy came a little later. It was introduced gradually.

---

[4]Uranium is an element with 92 protons in its nucleus and 92 electrons moving around the nucleus. These electrons determine the chemical behavior of uranium. The total number of particles, 235 in the nucleus, is made up of 92 protons and 143 neutrons. Uranium 238 has the same number of protons, the same number of electrons, and generally, the same properties. But it has 3 more neutrons and this makes a considerable difference. The rare isotope uranium 235 produces fission when it catches any neutron, fast or slow. The abundant isotope, uranium 238, gives fission only when bombarded by fast neutrons. The latter isotope is not highly fissionable; the former is. Highly fissionable nuclei are essential in atomic explosives and atomic energy.

I even claim considerable credit in the early decisions that eventually led to the Manhattan Project. I served on a very important occasion as a chauffeur for Leo Szilard. Szilard knew how to do almost everything in the world, but he couldn't drive a car. One particular thing he thought he did know how to do was to bring nuclear fission to the attention of President Roosevelt. That act was to be accomplished by writing a letter, having it signed by Einstein, giving it to one of Szilard's acquaintances who knew Roosevelt, and thereby getting the wheels into motion.

The letter was ready for Einstein's signature. He lived at the northern end of Long Island, and I drove Szilard there. Einstein received us in slippers, gave us some tea, looked at the letter and said, "Yes, yes, this would be the first time that man releases nuclear energy in a direct form rather than indirectly." By that time all physicists knew that the sun derives its energy from the atomic nucleus, and that we use nuclear energy all the time, but only after it has made its slow journey in the form of light through the interior of the sun, and from there has rushed to the surface of our planet in eight minutes.

Several months later President Roosevelt received the letter. He called the head of the Bureau of Standards in Washington, Dr. Lyman J. Briggs, and because Roosevelt told him to do something in response to the letter, Dr. Briggs called a meeting to decide what to do about fission. Szilard did indeed know how to get the atomic age on the road.

Several people were invited to the meeting, including Enrico Fermi and me. But Fermi would not go. He had gone to the Navy with the proposal of fission, had been thrown out, and had had enough of this nonsense. I was Fermi's friend, so I went to New York and tried to persuade him. He said, "I won't go, but I'll tell you what I would say if I did go, and you can say it for me."

So the meeting was held. The case for nuclear fission was presented. Then a representative of the Army spoke, a colonel whose name I have fortunately forgotten. This is what he said: "In Aberdeen we have a goat tethered to a stick with a ten-foot rope, and we have promised a big prize to anyone who can kill the goat with a death ray." (That was long before the days of modern death rays, the lasers.) "Nobody has claimed the prize yet. I don't believe any of this nonsense you scientists are talking, about death rays or atomic bombs. Besides, wars are not won by weapons, they are won by moral superiority." That would have sounded all right, except that it was November 1939, after Poland had been crushed by the moral superiority of Hitler and Stalin.

Soon it was my turn. I correctly stated that I was speaking for Fermi (I had advanced from chauffeur to messenger boy). Then I explained that one can begin to work on nuclear energy by slowing down in a particular substance—graphite—the neutrons emitted by uranium, and thus avoid their useless absorption by the abundant uranium 238. This graphite had to be very pure because any impurity would absorb the neutrons. After the neutrons are slowed they can impinge on the more readily fissionable uranium 235, and in this way we can carry on a chain reaction. Actually, much of this story had already been told, but I reminded those at the meeting that we had difficulties in getting really pure graphite. And then I said, "For the first year of this research we need six thousand dollars, mostly in order to buy the graphite." That is just what we got. My friends blamed me because the great enterprise of nuclear energy was to start with such a pittance; they haven't forgiven me yet, although I have meanwhile improved my ways, and learned to ask for bigger sums of money.

I would like to tell you of one other event that occurred before we really got going. The research developed slowly at the beginning, and I was undecided about whether I should be a bystander or a participant. I will tell you how I made up my mind, and when, and where, and why.

In those days I was a theoretical physicist, uncontaminated by any applications, let alone by anything akin to politics. I was much too interested in my own field. I lived in Washington, but I had not once visited the Capitol. I had not heard President Roosevelt speak even on the radio.

In the spring of 1940 it was announced that President Roosevelt would speak to a Pan American Scientific Congress in Washington, and as one of the professors at George Washington University I was invited. I did not intend to go. But the day before the scheduled speech, Hitler invaded Holland and Belgium. A statement in the press clearly indicated that the president would talk about that event. I had escaped from Germany, having come originally from Hungary. I had a reasonably clear idea, clear and horrible, of what conquest by the Nazis implied. So I decided to go.

I listened to the president. It was the first of his timetable speeches saying how long it would take to fly from this place to that to deliver a deadly attack. Since this was a Pan American Congress, the president talked about the right of small nations to exist and to determine their own way of life. Then he started to talk about the role of the scientist, who has been accused of inventing deadly weapons. He concluded: "If the scientists in the free countries will not make weapons to defend the freedom of their countries, then freedom will be lost."

I knew what he was talking about. I knew that he had read a certain letter six months earlier, signed by Einstein in my presence. I had the strange feeling that the president was speaking to me. When he was through, I was surprised to see that he had talked for only twenty minutes. But my mind was made up that I should be a participant, and I have not changed it since.

I will spare you the details of the work after it actually got going. I would like only to tell of one little incident in the middle of it, and then about the situation at the end of the war.

In the spring of 1939, Niels Bohr had argued with us against secrecy and against the possibility of making nuclear weapons. He went back to Denmark, almost landed in a concentration camp, escaped, and returned to the United States. He became part of the Manhattan Project at the Los Alamos, New Mexico laboratory. All the important people on the project had code names. His name was not Niels Bohr, but Nicholas Baker. We called him Uncle Nick,

Just before he arrived at Los Alamos I thought that now, for once, I can tell Niels Bohr that he was wrong (he often told me that I was wrong, and rightly so). In 1939 he had said, "You cannot separate the isotopes of uranium; you cannot make progress unless you turn your whole country into a huge factory."

I went to the laboratory the day he was to arrive, walked down the corridor, and saw Uncle Nick at a distance. When he saw me he came running, and before I had a chance to utter a word, he said, "Didn't I tell you that you couldn't do it unless you turned the whole country into a factory? And you went ahead and did just that!"

Our work had its results. We were going to test those results at Alamogordo, in the New Mexico desert, in the summer of 1945. For the last few months before the test I had a peculiar assignment. We had calculated the effects of the experiment. We *knew* approximately how big the explosion would be. But there were some who said, "Perhaps the explosion will run away. Perhaps we will blow up the world!"

According to the known laws of physics this could *not* happen. But could there be other laws of which we were ignorant? Could anyone dream up such laws, laws that could magnify beyond all expectations the effect of what we were going to do? Could one imagine natural laws that had

FIGURE 8-4 *Niels Bohr, the man who shaped modern physics and who was called "Uncle Nick" in Los Alamos.*

theretofore remained hidden? It became my job, and a very exciting one for me, to try to imagine such laws. I discussed it with others who worked with me, going really to everyone, to Niels Bohr, to Enrico Fermi, to everyone there—and the best physicists were there—and see what they could imagine. The only result we could reach, after long and careful thought, was that what we were going to do would be absolutely safe.

Some other things were not so safe. We had a little electric generator where we worked in Los Alamos that would go on the blink twice a week. The day before the Alamogordo test the lights went out, as usual. Trying to find my way home in the darkness, I bumped into an acquaintance, Bob Serber. That day we had received a memo from our director, J. Robert Oppenheimer, saying that we would have to be in Alamogordo well before dawn, and that we should be careful not to step on a rattlesnake. I asked Serber, "What will you do tomorrow about the rattlesnakes?" He said, "I'll take a bottle of whiskey." I then went into my usual speech, telling him how one could imagine that things might get out of control in this, that, or a third manner. But we had discussed these points repeatedly, and we could not see how, in actual fact, we could get into trouble. Then I asked him, "And what do you think about it?" There in the dark Bob thought for a moment, then said, "I'll take a second bottle of whiskey."

I watched that first explosion from a distance of twenty miles. It was just before dawn, barely starting to be light. We had been told to turn away, not to look toward the blast since it could hurt our eyes. Also, each person was given an oblong piece of the protective glass used in a welder's mask as an additional shield.

FIGURE 8-5 *Typical scenery near Los Alamos, New Mexico, during World War II. The physicists seem out of place.*

I wouldn't turn away. I wanted to look the beast in the eye. But having made all those calculations, I thought the blast might be rather bigger than expected. So I put on some suntan lotion. I put on a pair of dark glasses. I pulled on a pair of heavy gloves. With both hands I pressed the welder's glass against my face, making sure no stray light could penetrate around it. I then looked straight at the aim point.

The countdown came every ten seconds. After thirty it came steadily: thirty, twenty-nine, twenty-eight . . . , all the way down to five, and then there was silence. Nothing happened for an eternity, and I was sure that the experiment had not succeeded. Then I saw a very faint point of light that divided into three lobes, a ring with the rising fireball in the middle. My first reaction, which lasted less than a second, was, "Is that all?"

Then I remembered that my eyes were shielded by that extra glass. After a few seconds I very cautiously eased the welder's glass from its tight pressure against my face and peeked down at the sand beside me. Think of how it is when you open the curtains in a completely dark room and bright sunlight is pouring in at you—that is what I saw. Then I was impressed.

The first bomb, which destroyed Hiroshima, was exploded not many weeks later. It killed almost a hundred thousand people. I felt that this was terrible, that it should have been avoided. I had hoped that our first use would be a demonstration and that the Japanese would surrender after a demonstration. If we could have shown that the power of science could stop a most horrible war without killing a single person, I believe the world would be happier and safer, and the role of science would be better understood today.

These are my feelings. There is no way I can know whether I am right or not. I do know that President Truman was convinced he had to end the war surely and quickly. Europe was in ruins; starvation was to come the next winter. We had to be ready to help.

I have read and reread the history of Japan in those days between Hiroshima and the actual surrender. It is a most remarkable history. Before the bomb was dropped, of the six members of the war cabinet three were for surrender and three were against it. The dropping of the bomb did not change a single vote. The emperor intervened. He recorded a talk to the Japanese people, calling for surrender. This was an unconstitutional act on the part of the emperor, and the war faction staged a palace revolt. The leader of that faction, however, War Minister Korechika Anami, considered his highest duty to be to his emperor, who was God. He subdued the revolt of his own friends. Then, after the emperor's appeal for surrender—"We must endure the unendurable"—was heard and surrender assured, Anami went to his home and committed *hara-kiri*.

It is anybody's guess as to what would have happened had we demonstrated the bomb, and then used it only in case surrender was not forthcoming. I do not know, but my feelings are that it was wrong to use the bomb without first demonstrating it.

Let us move on to subsequent events in the atomic energy story.

In 1941, even before serious work had started on the fission explosion, Fermi had mentioned to me the possibility of fusion: to try to reproduce what is going on in the sun, not split nuclei but put them together. Might the atomic bomb be used as the match to start thermonuclear fire, fire that does not depend on neutrons but depends on heating material to temperatures higher than those found in the center of the sun, where the sun itself produces its energy?

It was a challenging question. Fermi asked me the question while I was working with him in New York. In two weeks I had the answer, and I told him as we went for a Sunday walk that it could not be done. I proved it to him, and he believed me.

In 1942, when the atomic project had really gotten going and I had full time responsibility to it, I found myself with nothing to do for a little while and with a very nice collaborator, Emil Konopinsky. I told him about that discussion with Fermi, and said, "Let's write down my proof so that people will not try to make the same foolish speculation over and over again." We tried to write it down, but the more we tried the more incomplete the proof seemed. In the end we came to the conclusion that maybe nuclei *could* be fused.

This was a beginning. We went through a number of reversals, a number of changes of plans. Ultimately I was absolutely convinced, not only that fusion could be accomplished, but that it *must* be tried. The hydrogen bomb appeared to be a definite possibility.

I remember when almost ten years later my friend Ernest Lawrence, who had encouraged me to work on the hydrogen bomb, called me, together with a number of other people, into his Berkeley office and asked for a report. I gave it to him. The same Bob Serber who would take two bottles of whiskey to Alamogordo was there, and he was absolutely opposed to the hydrogen bomb. When I was through with my recital Serber asked, "Edward, aren't these the same difficulties that you spoke about three years ago?" I said, "Yes, they are the same difficulties, but now I can describe them on the basis of much firmer knowledge."

One year later, in 1952, we were ready to fire the first fusion explosion. I can't tell you how that was to be done because it is still classified as secret information. I can tell you that I had left Los Alamos where this work was going on and had gone to California to help start the second weapons laboratory. I felt that while we were starting it I should not travel around, but should stick with the job. But I wanted to see that explosion.

This time there was no protective welder's glass and no desert. The experimental explosion was to be in the Pacific. I went down into the basement of the University of California geology building in Berkeley, to a seismograph that had a little light-point marking on photographic film. A tremor of that point would show when the shock wave, generated thousands of miles away on Eniwetok Island, reached Berkeley. I watched the light point but it would not stand still. Try to look at a point of light in the dark; it will dance before your eyes because your eyes are moving. I took a pencil and steadied it against the side of the apparatus; then I could see that the point of light, relative to the pencil tip, was steady.

FIGURE 8-6 *Ernest O. Lawrence, a driving force in work on U.S. atomic energy.*

At exactly the scheduled time I saw the light point move. It moved so slightly that I was not sure whether I just thought it moved or whether it actually had moved. So I stayed around for another ten minutes, lest I miss the real event; then I took the whole film and had it developed. There was the signal, just as predicted.

All of this of course was strictly secret. The explosion was conducted under the direction of the old laboratory at Los Alamos. I saw that they had succeeded. I felt I had to tell them. So I sent off a telegram which I hoped would not violate security. Its full text was, "It's a boy." That telegram was actually the first news they received of the success. The sound waves took twenty minutes to carry the message under tbe Pacific and arrive in Berkeley. It took another thirty minutes for me to send off the wire. But security officers at Eniwetok took a much longer time to clear any message to Los Alamos.

I want to tell of a sequel to the 1939 discussion about secrecy to which I referred earlier. Then I had argued that we should keep our work secret. The only one in our group who argued against secrecy was Niels Bohr. After the end of World War II, when Bohr was speaking of the cold war, which all of us saw coming, he said, "It would be reasonable to expect each side to use the weapons that it can handle best. The best weapon of a dictatorship is secrecy, but the best weapon of a democracy should be the weapon of openness."

I agreed with him at that time. I still do. I have seen our side try to keep secrets. I have seen the Russians, who can use the weapon of secrecy much better than we can, overtake us. I have seen us maintain leadership in the nonsecret field of fast computers, which are of great importance in peace and of equal importance in defense.

I argue for openness now. I hope that sometime we can have it, although I feel there is no hope at present that Russia would adopt a policy of openness even if we did. If we stop secrecy it would be a unilateral action. You might even call it a step toward unilateral disarmament. But it would not be to our disadvantage. We are, as Niels Bohr said, clumsy at handling the weapon of secrecy.

Opening up will not be easy, nor can it be quite complete. Day-to-day actions cannot be published. If the world knew where our submarines were at any moment they would no longer be useful. But if research is open it is more effective. The alchemists proved that secret science is no science. In the long run we cannot keep secrets, and we should not try.

We should do more than stop secrecy. We should use openness as a weapon. Every reasonable and gentle method should be employed to stop the spreading cancer of secrecy in the world. We should not help anyone who withholds long-term secrets. Of course, we cannot adopt such a policy unless we are ready to practice what we preach.

I dream of an open world where there are no national secrets. At first Russia would certainly not be a part of it. I hope the free nations would participate. If they did, the free world would be stronger and more united—demonstrating that openness is a weapon.

One great Russian scientist, Andrei Sakharov, has spoken out for openness. In doing so he has put his life in danger. Most Russian scientists do not dare to speak, but they feel the same way Sakharov does. If the free world adopts openness, we shall have allies behind the iron curtain. Then if openness proves effective in speeding up research, even the Russian leaders may change their minds.

I do not expect this dream to become a reality during my lifetime. But when it does come true, Russia will be less a police state, and we shall have made real progress toward a peaceful world.

It was openness that helped us get nuclear reactors underway in the Atoms for Peace Conference in 1955, so that we now have a potential source of energy that will last far longer than oil or even coal, and that could help to solve our present energy shortage.

In 1958 I had the great pleasure of winning a fight to liberate at least one field in nuclear energy, the search for controlled fusion, from the chains of secrecy. Controlled fusion is now an international enterprise in which the United States and Russia participate and exchange information. When I had the very pleasant job of raising that subject at the second Atoms for Peace Conference in Geneva, I also had the chance to talk about another project, one that holds a big promise for the future of atomic energy. It is called "Plowshare," the use of nuclear explosives for peaceful, constructive purposes.

Professor Emilianov, leader of the Russian delegation to that 1958 conference, immediately got to his feet and denounced this project as a dirty capitalistic trick introduced only to justify continued work on weapons. At a press conference a few hours later, a reporter asked, "Professor Emilianov, was it not a member of the Politburo who, in 1949, announced the first Russian nuclear explosion as devoted to peace, to the irrigation of deserts, the moving of mountains, the improvement of men's lives?" Emilianov replied, "Yes, but that was a politician speaking, and we Russian scientists are never influenced by politicians."

We should note the fact that now the Russians, who are not limited as we are by extremists parading under the banner of environmentalists, have moved ahead of us in Plowshare, and they have said so very clearly and plainly. Don't expect me to explain this development. Maybe politicians have an influence in Russia after all. Whatever the reason, the Russians surely make up in flexibility what they may lack in consistency.

I know of no invention that could not be used in peace and war alike. I am confident that nuclear explosives will, in the long run, be used only for peaceful purposes. Another, older story, that of nylon, serves for comparison. Nylon is a weapon in the battle of the sexes and as such is considered to be a peaceful instrument. But when DuPont succeeded in making nylon, women did not get any stockings. We were in the middle of World War II and all the nylon then went into parachutes.

What science does is to teach something and, incidentally, to increase our control over nature. Once we have that power, it can be used or it can be misused. I hope that nuclear energy will never be misused again.

Several questions about the atomic bomb and the hydrogen bomb are troubling to many. I believe that these questions deserve the fullest possible answers.

A question frequently raised is: Should we expect the development of even more destructive weapons? There is in all of us a tendency to exaggerate, and it is easy to see why many people have believed that the end of the world is approaching. The A-bomb increased a thousandfold the destructive power of which man is capable; the H-bomb turned out to be a thousand times more powerful than the A-bomb. Where will this end?

The fact is that in the past twenty years, escalation of explosive energy has stopped. We have more sophisticated weapons, more easily delivered weapons, even weapons that can destroy a nuclear bomb approaching at a speed of many miles per second. But neither we nor Russia nor any other nation has produced much bigger bangs.

But even without bigger weapons we do have more nuclear explosives. Could the human race survive an atomic war?

In a time of incredibly rapid change any prediction becomes dubious. But in all of known history, the capacity for inflicting damage has never been what set the limit on destruction. It has always been the intentions of man that led to the terrors of war and that set the limit where terror had to stop.

We are torn between two extremes. Sometimes we believe we are better than our ancestors, and that there will be no more war. Sometimes we believe that, for the first time in history, we shall not know how to stop, but will bring history itself to a tragic end. I believe that neither the hope nor the fear is fully justified. I do believe that the atomic age has brought about a new situation. It has made the world more interdependent; we are all neighbors.

The atomic age has also compressed our time scale. What worries me is not so much the size of the trouble we may face, but the speed with which trouble may arise. Now as never before we must think ahead. There may not be time to correct mistakes should a worldwide conflict erupt.

I have not answered the question to the satisfaction of anybody. The world is not safe. It never has been safe.

In the minds of many this lack of safety seems to be caused by scientists. They believe that the scientist must take responsibility for the dangerous tools he has constructed. I do not agree. Some people overestimate the responsibility of the scientist. They also overestimate his foresight and his power. The role of the scientist is and must be modest.

The scientist's prime responsibility is scientific knowledge. If he does not succeed in finding the laws of nature, nobody will. Knowledge is different from power. It has an intellectual value that will outlast anyone's power.

The scientist's second responsibility is to turn knowledge into practical applications. The result is technology, and technology has a great humanizing influence. It may make us better or worse, but it will undoubtedly increase the difference between ourselves and the rest of the living world. This means that we will be even more human. Technology transcends national boundaries; it will not be lost.

Finally the scientist should explain his findings so that everyone can understand. He must explain what he has understood and what he has constructed. The nonscientific world then can choose to use or misuse the new knowledge.

If a scientist went farther, if he made decisions on his own, he would contradict the basic postulates of democracy. I believe in decisions by the people and for the people. Anything else is elitism. If scientists consider themselves an elite with responsibility for the world resting on their shoulders, they will no longer be scientists.

I am sometimes asked if I am sorry that I worked on nuclear weapons. I certainly am not. I had the opportunity to participate in one of the most fantastic adventures that a scientist can have. I had the chance to help put powerful tools into the hands of my fellow citizens. I feel confident that, on the whole, these tools will be used properly.

The atomic age may be an age of fear, but it is also an age of hope. We have seen that the discovery of nuclear energy was delayed by some remarkable accidents. The eventual development of atomic weapons was unavoidable. In the wrong hands these weapons would have become instruments of oppression. You can perceive from my recollections of the early days of atomic energy that I have no regrets about my part in it. I am happy to have been among those who established man's control over the atomic nucleus. I cannot be responsible for all the consequences of my actions. No one ever is. To believe in such a comprehensive responsibility would be akin to believing in one's own omnipotence.

The atomic adventure was not the first, nor the last, nor the greatest adventure in history. But it was great and it was inevitable.

The reader may conclude that, in this perilous world, I remain an optimist. Let me give a couple of definitions. A pessimist is a person who is always right but does not get any pleasure out of it; an optimist is one who imagines that the future is uncertain. The two definitions do not exclude each other.

Indeed, I am an incurable optimist, one who does not desire to be cured. I imagine that the future is uncertain. And because it is uncertain, I am determined to do something about it.

CHAPTER 9

THE REACTOR SAFEGUARD COMMITTEE

In which is described the situation of a distant, simpler past, when the difficult problems of reactor safety were reviewed by experts without interference from concerned but poorly informed scientists, environmentalists, and the daily press, and also before the urgency of an energy shortage or the investment of billions of dollars.

Hiroshima was a traumatic shock to the world. What had been right, what wrong, was unclear. Two points seemed obvious. One was that atomic power would contribute to a different future. The other was that peaceful use of atomic power appeared desirable. It seemed as if a lot of energy could be obtained for almost nothing.

But this is not the way in which the world is put together. We do not get something for nothing, nor even a lot for very little. Beneficial use of the atom, however, is within reasonable reach. From postwar history we can perceive that real advantages are available from this new energy source—for a price.

Shortly after World War II it was decided that nuclear energy should be under civilian rather than military control. This decision focused attention on a principal peaceful use of atomic energy, electricity produced by fission reactors.

In 1948, early in its existence, the new Atomic Energy Commission (AEC) established the Advisory Committee on Reactor Safeguards. I became the committee's first chairman. The committee was composed of several scientists and a single engineer. The engineer resigned after the first meeting because, he said, safeguards could be established only on the basis of experience, not on theory. Industrial reactors were too new; there were no statistics. Therefore, in his opinion, safeguards could not be established.[1]

It is understandable that completely novel structures should appear frightening to an engineer with limited nuclear experience. He was frightened by Hiroshima. And he was not the only one. A nuclear reactor, however, cannot be turned into an atomic bomb. It is not the presence of great amounts of energy that is essential to an atomic explosion; it is the fast release of this energy. An intricate special design is necessary to make a fast release of energy possible. This design is lacking in reactors. A reactor can release some excess energy, but the reactor shuts itself down long before as much energy is released as would result from a common chemical explosion.

In principle a nuclear explosive is a rather simple device. The main point was recognized by Szilard as early as 1935. In more than a critical

[1] A quarter of a century later some insurance companies still hold to this position—that reactors cannot be regularly insured because there have not been enough accidents to justify statistics. Paradoxically, a perfect record can preclude regular insurance. To the layman, however, lack of regular insurance means lack of safety. Fortunately, the participation of insurance companies is increasing. They have noticed that nuclear reactors are indeed safe.

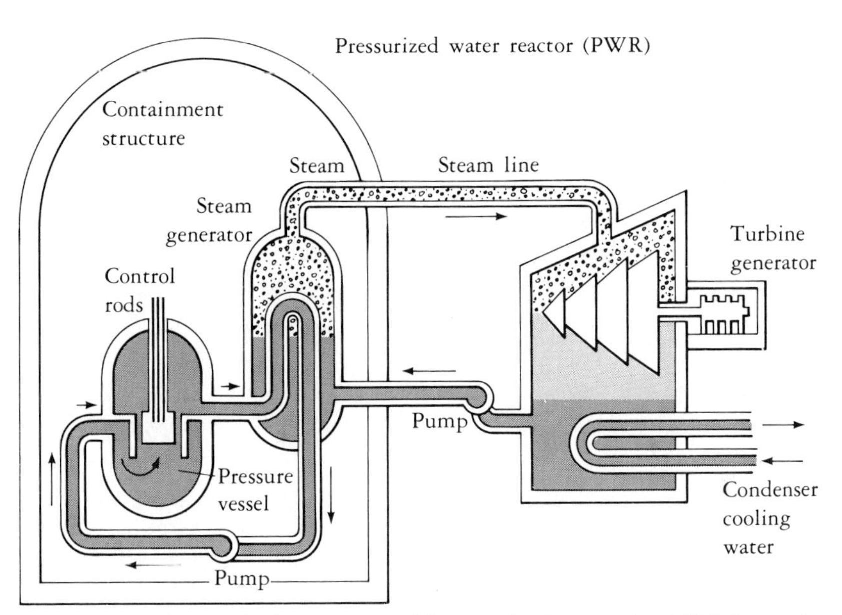

FIGURE 9-1 *A pressurized water reactor. The water in the reactor (at left) is kept under pressure so that it does not turn into steam at high temperatures. Vertical control rods can be slid in or out between the fuel rods at the core. The containment structure is about 20 feet in diameter and 45 feet high and weighs several hundred tons. Outside the reactor the pressurized water exchanges heat with water in another circuit. The steam produced drives a turbine that generates electricity.*

mass (the technical term for a sufficiently big mass) of fissionable material, a single neutron will produce a lot of energy and more than one neutron. These neutrons, in turn, produce more energy and more neutrons. After the number has doubled eighty times, practically the whole mass is involved in a violent reaction. This eightyfold repetition takes only a fraction of a microsecond.[2] If it took longer the mass of fissionable material would move apart more slowly and no big explosion could result. Since some neutrons can be found almost any place (produced by cosmic rays, for instance), it is essential that enough of the explosive fissionable substance be brought together in a very short time. This is the difficulty in producing a nuclear explosion.

A nuclear power reactor works quite differently. Here the emphasis is on releasing energy in a slow and controlled fashion. The usual procedure

[2] A microsecond is one-millionth of a second.

is to slow the neutrons down by repeated collisions with lighter nuclei such as carbon, hydrogen nuclei (protons), or heavy hydrogen nuclei (deuterons). This process slows the reaction. Further slowing occurs because a reactor is never allowed to become more than barely critical, that is, it never replaces a neutron by much more than one neutron. In steady operation the reactor is just critical, and the replacement goes one for one. Finally, some neutrons are generated with a delay of many seconds. The fine tuning of the reactor results from the insertion of control rods made of material that absorbs neutrons. The insertion of many or big control rods is the means of shutting down a reactor. Thus the reactor is rather easily regulated, and the fast reaction needed for an explosion is impossible.

This does not mean that reactors are completely without risk. Great amounts of radioactive materials accumulate in a power reactor. Indeed, in each fission process two unstable radioactive nuclei are produced. In a serious reactor malfunction some of these radioactive substances may escape. Clearly, too much radioactivity is harmful. A big dose kills. The dangerous levels have been firmly established. In a major reactor accident, people who are downwind from the reactor could be exposed to lethal radiation. The danger would extend for many miles. Thousands of people could be affected, and an upper limit is not easily established.

It is incorrect to say that in 1948 we had had no experience. Three reactors at Hanford, Washington, had been operating safely for years. In Los Alamos its wartime director, J. Robert Oppenheimer, predicted that in the laboratory there would be nuclear accidents and people would die. Fortunately, this did not happen. Car accidents occurred on the road winding up the mesa; people were injured while riding horses. But the atomic materials were handled with great care and with complete safety.

After the war many people, including Oppenheimer, left Los Alamos. Some who remained became overconfident. Twice, careless experimentation resulted in people being hurt. They were hurt needlessly; a minimum of caution practiced by all would have prevented these incidents. But two people were killed and more were injured. In both accidents people were experimenting with arrangements similar to nuclear bombs, not similar to nuclear reactors.

The Reactor Safeguard Committee knew all of this in 1948. We decided that in nuclear reactors incomparably greater precautions would have to be taken than were exercised in the postwar Los Alamos experiments. We realized that a single accident in an industrial nuclear reactor could wreck hopes for the peaceful atom.

Our purpose was to prevent any future loss of life, particularly in connection with regular industrial operations, which we hoped would become possible in the future. There was no alternative to extreme caution. We acquired a peculiar reputation among the community of nuclear scientists and nuclear engineers; our committee was dubbed the "Committee for Reactor Prevention." If the champion of seatbelts had been known at that time I undoubtedly would have been called a "mini-Nader."

We set about taking practical steps. The first was to break down excessive secrecy. According to law in 1948 and the years following, reactors were secret. More than that, information about them was compartmentalized. One group planning reactors did not know what another group was doing. This compartmentalization was not part of the law, and we managed to abolish it. Our discussions and hearings were opened to all people with legitimate concerns in the construction of reactors, and, in fact, planners and would-be planners of nuclear reactors came to attend for the same reason students like to attend other students' examinations. Everyone is interested in finding out ahead of time what kind of questions he will have to answer.

Personally, I went still farther. I advocated abolishing secrecy about reactors and revealing knowledge and potential dangers to the public. This was accomplished, for the most part, in 1955, when I was no longer on the Reactor Safeguard Committee. Today discussion has become not only open, but also widespread and often exaggerated. Exaggeration and sensationalism seem to be among the prices we pay for democracy.

We adopted some further steps that were quite straightforward. Nobody understands complex equipment as well as its designers. Therefore, we required that every proposal for a reactor be accompanied by a special evaluation that provided answers to two questions: What is the maximum credible accident? What are the consequences of a maximum credible accident?

Of course, the answer to the second question was always favorable. A designer of a reactor would never describe events that could lead to unacceptable consequences. It was nevertheless important that the questions be asked and answered. The designers were at least forced to evaluate in detail what might follow from their actions.

The first question, concerning the maximum credible accident, was not so easily answered. On that score we had to refuse licenses on two occasions. I shall describe one of these rejected proposals because it illuminates the improbable circumstances that might actually take place and cause an accident.

The maximum credible accident in this case was estimated to be smaller than an accident that had actually occurred near Chicago at the Argonne National Laboratory. This accident had taken place just before the meeting about licensing an experimental reactor that was to serve as a model of an engine in a nuclear-propelled aircraft.[3] The accident happened in an experimental facility of the laboratory and caused radiation sickness in four persons. Fortunately, all of them survived. It should be emphasized that no similar accident has ever occurred in the operation of industrial nuclear reactors. The mistakes in the experimental stages contributed to improved safety in industrial development.

The Argonne accident occurred because four bad mistakes were made by the person in charge. First, he was determined to complete an experiment even though he had not understood the behavior of the reactor immediately before the accident occurred. Second, with three other persons he violated the rule against entering the reactor chamber while there was water in the reactor. Water was needed for the reactor to function. The reactor was known to be safe when that water was absent, and was equipped with interlocks that prevented anyone from entering until no water remained within it. These interlocks were actually removed, in a conscious violation of this necessary safeguard. Third, after the people entered they not only looked, but they proceeded to change the configuration of the reactor. Finally, after other changes were made, the control rod was lifted out, and not by carefully controlled machinery but by human hands. To a reactor operator with any education, this action was truly incredible. (Still, according to our use of words, we have to include it among the "credible" accidents because it did occur.) A glow was seen, and the four people in the chamber hurried out. This saved their lives. The mechanism worked, which guaranteed that in case of an accident, water would flow out of the reactor and shut it down. Outside the reactor chamber there was no radioactivity and no trouble.

Since the proposal for a practically identical new reactor stated that radiation could not be released in quantity such as described above, we had to deny a license. Later a version with improved safety features was submitted and subsequently licensed.

It is virtually impossible to enumerate all the things that might conceivably go wrong with a reactor. Even if we discount human folly, the

---

[3]Work on the nuclear-propelled aircraft was later justly terminated. While the fuel would have been light, the advantage was cancelled out by the heavy shielding needed to protect the crew. Furthermore, a crash of such an aircraft in a populated area would have been a dreadful disaster.

variety of possible accidents and malfunctions remains too great. Each type of reactor has its special weakness. Good engineers know that, like everything else in the world, every piece of machinery has at least one Achilles heel. But I will give you an idea of the variety of components or functions in which things may go wrong. One of the most basic risks is a *criticality accident.* A reactor operates at a strict break-even point. There must be a balance between neutrons emitted and neutrons absorbed. Fortunately, approximately one percent of the neutrons are delayed about a minute. It is generally easy to keep the average balance exactly where it should be by using automatic devices and proper human control. But if for any reason this balance is upset, for instance by removal of a control rod, rapid changes may occur. The changes may be self-limiting. The consequence of too many neutrons will be even more neutrons and more energy. Usually more neutrons and more energy will eventually shut down the reactor and trouble can be avoided. A runaway of neutrons and energy is called a criticality accident.

As stated previously, a criticality accident can never turn a reactor into an atomic bomb; the processes in a reactor are much too slow. But a criticality accident can cause high temperatures in specific parts of a reactor that may lead to the release of radioactivity—even in massive amounts—accumulated earlier in the reactor operation.

One possibility is melting of the fuel elements. Indeed, this was our first worry. In modern parlance this is called a *meltdown.*

Meltdown of the fuel elements is by no means the only danger. Another possibility is that the metal jacket of the fuel elements may become very hot and then react chemically with adjacent cooling water (the committee discussed an aluminum jacket as an example). Remarkably enough, this chemical energy is apt to exceed the limited nuclear energy released in a criticality accident. When our committee first raised this possible problem, it was dismissed as contrary to all chemical experience. Later it was shown that if the metal jacket is heated and melted with great rapidity the reaction *can* take place,[4] and might result in release of radioactivity that had been deposited during normal operation.

Discussions and experiments, however, are never sufficient to insure safety. They encompass only a small part of reality. The total engineering

[4]Aluminum and other metals owe their usually inert, harmless behavior to an oxide layer that adheres closely to the surface. In rapid melting under water this layer is destroyed, and a violent reaction with the water may occur.

reality is the reactor itself. According to the wisdom of the centuries of the Industrial Revolution, the final guide had to be experience in actual use. In our Advisory Committee on Safeguards we realized that we could not permit even a single big nuclear accident that might cost many human lives. We wanted to reduce the probability of such an accident to zero, and that appeared almost hopeless.

We therefore proposed a radical remedy: to establish, in a deserted location, an experimental facility where we could cause nuclear reactors to malfunction in practically any manner imaginable. Needless to say, this suggestion was not greeted with cries of joy by tbe AEC, but the weight of arguments prevailed. The experimental site was established in Idaho.[5]

No one should be misled into believing that the arranged accidents at our Idaho site represented, or could ever represent, a final solution to the problem of reactor safety. The difference between an experiment in Idaho and a real reactor accident is that in the latter a lot of radioactivity would accumulate, a condition very difficult to reproduce. Real cases never can be mocked-up in a completely faithful manner. There must be three solid bases for confidence: careful calculation, observation of the contrived malfunctions, and thorough discussion correlating the two. Absolute answers are never available, but excellent approximations of reality have been achieved.

How, for instance, could we ever hope to mock-up an earthquake? But earthquakes certainly cannot be disregarded. We had to be satisfied with doing the best we could. The recommendation was made and accepted that any substantial earth tremor should automatically shut down a reactor. That would significantly improve reactor safety. We must also remember that a reactor is a solid structure to which the shaking of an earthquake is apt to do less damage than that done to the hollow buildings in which we live. Furthermore, from the very beginning, the location of reactors with respect to earthquake faults has been a most important consideration.

The earthquake hazard triggered an amusing incident in the Safeguard Committee discussions. A reactor planned for Brookhaven, Long Island, not far from New York City, was to consist of two slabs

---

[5]Some time later, at a different Idaho site, an accident at an experimental reactor killed three people. It occurred at an experimental facility not connected with the industrial reactors under discussion here. To this day we have not found out the reason for that accident. Was it extreme folly, or a suicidal action? There were no survivors. Experimental reactors are subject to fewer controls than industrial reactors are.

separated by an air vent through which heat could be removed. I worried that in a violent earthquake the two slabs could slide together, and the reactor might become critical. So we called in an earthquake expert.

Some of the best earthquake experts in the United States are found among members of the Jesuit order. Accordingly, we invited a learned Father from Fordham University to be a consultant. He had no security clearance, and had to be brought into the AEC building under guard. In the committee room he was seated in a deep armchair in which his small body almost vanished.

Members of our committee fired questions at our expert for half an hour. At length we ran out of questions, though he certainly did not run out of answers. When this had become obvious, he looked each of us, in turn, straight in the eye, taking me on last. Then he straightened, seemed to grow in stature, and addressed me. "Dr. Teller, I can assure you on the highest authority that there will be no major earthquake on Long Island for the next fifty years." It was hard, but no committee member cracked a smile until the door had closed behind the Jesuit Father who had brought word to us from the highest authority.

There was one argument in the committee which, to my great regret, I lost. I argued for underground siting of nuclear plants, but that would have required careful engineering analysis not then available. Instead, it was decided to build sturdy containment domes around reactors. There is excellent reason to believe that the dome would contain dangerous radioactivity even if such radioactivity should escape the reactor. I believed then and still believe that a reactor under two hundred feet of packed earth would be even safer. Perhaps, eventually, that is how our reactors will be built. New engineering and new safety analyses would, of course, be required.

The above includes only a scant portion of the discussions and worries we had and the precautions we took during the three or four years I was active on the Advisory Committee on Reactor Safeguards. Traditions established during those years have been carried on. The committee became an institution and its findings were opened to the public. Its work continues. Big segments of the public disregard or underestimate this work, which has proceeded for nearly thirty years. Committee members were in general not interested in the financial success of reactors. The heavy responsibility of creating reliable safeguards was felt by all of them.

I do not believe that any branch of technology has ever had as thorough, as expert, or as detailed a safety review as industrial reactors had even before the first one was erected.

FIGURE 9-2 *Seen at Oak Ridge National Laboratory, one of the main centers of reactor research, are the Periodic Chart of the Atoms and some of its users. From left: laboratory director Alvin Weinberg, Eugene Wigner, Clarence Larson, the author, and James C. Bresee.*

This chapter would be incomplete without mention of one serious accident, which occurred in England in the late fifties at the Windscale reactor.

During the war Eugene Wigner, who is perhaps the greatest of reactor experts, worried about the energy deposited in graphite by radiation. Graphite is an important component of many reactors. The energy deposited is gradually released without harm if the graphite is fairly hot. But if it is at a lower temperature and local heating should occur, the energy release could raise the temperature, which would further increase the release of energy. This happened in the Windscale reactor; it actually split open and the graphite started to burn.

At that point the operators made the right decision. They shut down the reactor. They did not turn on the water hose, which probably would have resulted in a steam explosion that could have torn the reactor open more widely and would have liberated great quantities of radioactivity produced earlier. The fire had to be smothered by carbon dioxide. Unfortunately, and inexcusably, sufficient carbon dioxide was not at hand. The British let the fire burn, brought in carbon dioxide as fast as possible, and put out the fire. Some radioactivity did escape, but damage was limited.

To the best of my knowledge, this is the closest the free world ever came to a serious reactor accident. Today reactor safety and the all-important question of whether new reactors should be constructed are at the center of public interest. What has been described in this chapter is a prelude; it illustrates that worries about reactors were not born yesterday.

Now we should turn to the question of what should be done at this time, when the energy problem has become of overriding importance. This question cannot be settled without thorough discussion of the many arguments that have been made both for and against these remarkable big instruments. To prophets of the apocalypse, reactors are precursors of the end of the world. In the opinion of others they are a major and necessary part of solving the energy crisis.

CHAPTER 10

REACTOR SAFETY AND THE ANTINUCLEAR MOVEMENT

In which opponents of nuclear reactors appear in the role of that enormous rock Sysiphus tried to roll to the top of the mountain in the Underworld. At great effort, Sysiphus always succeeded; but the rock always rolled down the hill, so he had to start afresh. (Fortunately, the reader will have to read this chapter only once.)

Anything that is new and strange is frightening. A new technology, developed in strictest secrecy, was first known to the world when a hundred thousand Japanese were killed and a great war was suddenly ended. Fear was unavoidable.

Fear was delayed, however. In the first years after World War II, the irrational aspects of fear did not show up. Relief at having the war over was enormous. Hope for a permanently peaceful world seemed real to many people. The need to reconstruct a world left in shambles was immediate.

Soon there were signals of danger: trouble in China, the division of Europe, failure of the Baruch plan for internationalization of atomic control. Work on atomic energy continued in a relatively inconspicuous fashion, and the public generally seemed to accept control of the atomic nucleus as one more strange fact in an increasingly strange world where peace seemed to supplant the immediacy of dangers.

But complacency was unrealistic. The explosion of the first Russian atomic bomb in 1949, and development of the hydrogen bomb in 1952, could not be ignored. President Eisenhower decided that a new encouragement, a renewed impetus toward international cooperation could come from the peaceful atom. He proposed that controlled fission, for producing energy from atomic sources, be used to strengthen the fabric of peace. The idea for this development actually came from the president's friend Lewis Strauss, head of the American Atomic Energy Commission, who urged an "Atoms for Peace" conference.

Proposed by the United States, the first International Atoms for Peace Conference was held in Geneva in 1955. The Kremlin answered its initial invitation with a resounding *"nyet."* But when it became clear that there was going to be a party, with them or without, the Russians came. Happily, their scientists made genuine contributions and enjoyed the conference. They were not the persons who had refused the first invitation.

That conference substituted euphoria for anxiety. The age of the peaceful atom had arrived.

The good life depends upon ample energy if that good life is to be achieved by the means realized through the Industrial Revolution. Toil was reduced by control of the forces of nature. The elimination of suffering from poverty by reasoned use of technology came closer to reality in the United States than in any other society. If a similar result could be achieved for a world recovering from a terrible war, if the age-old hardships in the underdeveloped portion of the world could be brought to an

FIGURE 10-1 *Uranium fuel is made into small cylindrical pellets about the size of the end of a little finger. The heat energy of one pellet equals that of one ton of coal. Before entering the reactor, fuel elements are weakly radioactive and can be held as shown in the sketch. (George Bing.)*

end, then perhaps the exploitation of man by man could come to an end. The American Dream could become a reality throughout the world. But ample energy was one condition for achieving all of this.

Today more people realize that curing the world's ills is not so simple. Energy alone will not do it; neither will any straightforward change in social order. Whenever we have been too optimistic we have stumbled. Nonetheless, there was a point to that optimism. An immediate benefit seemed within reach in the 1950s. When trouble on the Suez Canal threatened the flow of oil in 1956, many anticipated a ready solution. They believed that nuclear reactors would deliver energy soon and in great quantities.

Their hopes were disappointed. Nuclear reactors proved to be impractical at that time because of their cost. It was cheaper to ship oil in big tankers even though they could not navigate through Suez. The cost of the long voyage around the Cape of Good Hope was more than compensated by savings realized from the great capacity of the vessels.

America's national laboratories, particularly at Argonne near Chicago and at Oak Ridge in Tennessee, worked in secrecy. Scientists there had developed the principles of the great nuclear reactors but had not yet progressed to effective, economical engineering application.[1]

After 1955, when schools of nuclear engineering were established, private enterprise took over. Costs were gradually reduced, and reactors became practical. By 1967, twelve years after the first Atoms for Peace Conference, they appeared to be competitive in price.

And none too soon! The supply of petroleum, which had all but monopolized the energy market, no longer seemed inexhaustible for the future. In the United States a movement for protecting and improving the environment had made a belated but necessary start. A new supply of energy, plentiful, clean, and economical though not cheap, would have been enthusiastically welcomed. Yet this time also marked the beginning of serious difficulties for development of that very energy supply.

Fuel for nuclear reactors, including separation of the active isotope uranium 235, was inexpensive. The greatest cost was—and still is—in capital investment.[2] In the late 1960s and early 1970s, a sharp rise in interest rates worked to the disadvantage of nuclear industries, which needed long-range capital investment for an energy yield that would be realized only after the elapse of several years.

The oil embargo of 1973, with a fourfold increase in oil price, highlighted the desirability of nuclear reactors. But after widespread planning for reactors there arose a chorus of objections to these plans. Objections proliferated.

These objections were not an outgrowth of nuclear accidents, although some accidents had occurred. Anyone who imagines he has found a foolproof system is apt to learn that the fool is bigger than the proof. Industrial reactor accidents resulted in the loss of many millions of dollars.

---

[1]Admiral Hyman Rickover adapted this work to nuclear submarines—well-engineered, extremely safe, but expensive instruments of warfare. (The first, the *Nautilus,* was commissioned in September 1954.) In the course of time nuclear submarines may become cheap enough to haul oil under the Arctic ice cover to consumers all over the world.

[2]This point should be remembered by advocates of novel energy sources including abundant solar energy. At present a solar electricity generating plant would require investment at least five times greater than the capital investment for a nuclear reactor of comparable average generating potential.

But not a single person was killed, nor did the health of a single individual suffer because the reactors were nuclear. As stated in Chapter 9, no one has been injured as a result of the operation of industrial nuclear reactors. To prevent misunderstanding I reiterate: the accidents referred to in that preceding chapter occurred in experimental reactors or weapons laboratories; they have no parallel in the nuclear industry. The ambitious aim for achieving operation of big nuclear plants without human suffering has been attained. The protests were not based on actual lack of safety.

There is a rhythm in industrial development and in public criticism of that development that may result in an unfortunate countermarch. During the expensive initial phases of planning, research, and development, there is little criticism. When plans have matured, much effort and investment have been expended, and the potential product is clearly perceived, then public protest can become clamorous and obstructive. It may develop into an insurmountable obstacle.

The cause is lack of foresight, and responsibility for that should be shared. In the case of reactor development, part of the responsibility must be attributed to excessive concentration of authority. Originally the same office was responsible both for new development and for licensing, that is, giving the "go-ahead" signal. Thus, the same person was given authority for creating, and for pronouncing the creation as good. (The Bible reserves this dual role to God.)

This arrangement has been modified in a realignment of authority for development and licensing. The Energy Research and Development Administration (ERDA) was made responsible for new development,[3] while the Nuclear Regulatory Commission (NRC) makes decisions on licensing. The NRC was first headed by Bill Anders, an engineer astronaut who staked his life on the faultless functioning of more than 100,000 parts in the rockets that carried him around the moon and back. Unquestionably, he knows about safety.

More responsibility should perhaps be laid to the critics who voiced their criticism belatedly, and then pursued it in many cases with broad and unsubstantiated arguments rather than specific reasons related to engineering.

---

[3]Now replaced, further, by the Department of Energy (DOE). If in doubt, reorganize!

The antinuclear movement has become widespread and powerful. The only way to meet and understand it is to discuss in detail each argument that is raised against reactors. This may not convince professional objectors, but it is necessary for the general public, whose vital interest is at stake.

The first question is: What is the possibility of a major reactor accident? The industrial reactor health record so far is perfect. But can reactors remain as safe in the future? Might the law of probabilities turn against us? Can we continue to guard against human error?

Reactor safety results from multiple safeguards. Every possibility of malfunction is prevented by three or four or even five independent measures.[4] An impairment of any of these safety measures is considered an "accident" and the deficiency is repaired, regardless of cost.

The actual probability of an accident, particularly a major one, has been evaluated in a careful engineering study by Professor Norman C. Rasmussen and his associates at Massachusetts Institute of Technology. Their conclusion is that the probability of a big accident is less than one in a million. To put it in more familiar terms, a person is not likely to be killed by lightning, but it does happen. Nuclear accidents in industrial reactors are far less probable. A remarkable, and more quantitative comparison was made in the Rasmussen report: The probability of a single person being killed by a meteorite is similar to that of his being killed by a reactor; that ten persons should be killed by a meteorite or a reactor is far less likely but the probability for each is similar; that one hundred should be killed by either is again equally improbable; that one thousand people should be wiped out is so unlikely as to defy imagination, but the probabilities are essentially the same for meteorite or reactor.

Rasmussen's report was prepared meticulously. Each type of possible malfunction was evaluated on the basis of detailed engineering experience. Yet the report has been criticized and the criticism is to some extent justified. One may argue that the Rasmussen report is too optimistic; one may also argue that it is not optimistic enough.

It is extraordinarily difficult to estimate the probability of excessively improbable events. The greatest danger remains: something may have

---

[4]One popular example of a pessimist is a man who wears both a belt and suspenders. In case of nuclear reactors we added even more safeguards. Thus one may say we are superpessimistic. It pays.

been forgotten. It is impossible to estimate the limits of our own imaginations.

On the other hand, the multiple reactor safeguards give us a special benefit that is not calculated and is virtually impossible to calculate. It is extremely unlikely that all the safeguards would fail at the same time, and the safety from this circumstance is taken into account. But the failure of a single safeguard leads to long-term improvements in safety. Any such failure results in substantial monetary loss, since the reactor is not permitted to operate unless all safeguards are in good condition. To avoid financial damage, the safeguard is improved. Thus, year after year, reactors become more safe. There is a dent in the pocketbooks, but not in the people. Rasmussen could not take such future improvements into account.

One example of improvement through accident occurred at the famous Browns Ferry reactor in Alabama in the spring of 1975. A candle was used in testing for leaks in a supposedly airtight part of the reactor. The flame was sucked into an inaccessible part which contained complex wiring. An electrical fire started. The reactor was shut down, but because of mistakes the fire was not put out for several hours and a number of safety devices were knocked out. The financial loss exceeded $120 million, but damage to health was zero. A danger point for health damage was not even approached.

Of course no hundred-million-dollar candle will ever be similarly used again. Furthermore, we have learned that wiring which serves independent safeguards must be better separated. The accident was a demonstration of the health safety inherent in reactors. At the same time the accident led to improvements in reactor construction. And that is most important.

A specific imaginary accident has played a great role in the reactor debate. William D. Kendall, a physicist not previously concerned with reactors, rediscovered the possibility of an "emergency core cooling accident." The complicated name describes a complicated situation. It is characteristic of the discussion in its present state that Kendall's objection is concerned with a double failure: a rapid loss of cooling water due to complete rupture of piping, and a subsequent failure of the standby system for delivering the coolant water held in readiness for such an emergency.

Despite the complexity of the hypothetical problem, it is worthwhile for the public to understand the merits of this debate. It can illustrate the way reactor safety is discussed today.

First, it should be emphasized that the functioning of the emergency system would be important only in a very peculiar situation: If the proper primary coolant is lost rapidly, the nuclear fuel may become so hot that the emergency coolant cannot enter; it may vaporize on its way in. If that should happen, the nuclear fuel may melt and radioactive products may be released in great quantity. But they would not be released to the outside. The whole reactor is surrounded by a strong container. Even in the worst case envisioned, no one would be hurt.

Two improvements were introduced after the possibility of failure of both the primary and secondary cooling systems had been emphasized. One was to decrease the diameter of the rods containing the fuel. This reduces the heat that the coolant must carry away after the reactor is shut down, because the temperature near the center of a big rod must be quite high in order to create the necessary heat flow from the rod's axis to its surface. Heat energy within the nuclear fuels can be reduced considerably by using smaller rods. It must be realized that use of these small rods will not of itself suffice; after a reactor has been shut down, the nuclear fuel is strongly radioactive and continues to generate more energy—a small amount compared with the energy generated during operation, but still enough to cause melting. With smaller rods, it will take a longer time to reach the melting point.

Of course, the radioactive "afterheat" will not go on forever. The longer we wait, the less the danger. A second after shutdown considerable energy is produced, requiring much coolant in a short time. An hour after shutdown the rate of delayed heat production is reduced ten thousandfold. Afterheat continues to diminish with the passage of time.

The second and perhaps more important improvement in the emergency core cooling system is that in an emergency situation water is injected at higher pressure. Thus it can overcome the pressure of the vapor that may result when the coolant hits the hot rods.

It is indeed fortunate that decades ago in Idaho we established the system that permits simulation of practically any accident, including the failure of the emergency core cooling system. Such simulations have been carried out, but we can hardly expect a simulation to reproduce an actual accident with complete accuracy. Furthermore, it is impossible to reproduce all the various conditions under which accidents can be imagined. To ask for a complete safety test in the emergency core cooling system is similar to demanding that automobile accidents be simulated, evaluated, and then avoided under all conditions that can occur in actual driving.

One argument made by opponents of nuclear reactors is that, while car accidents are terrible, nuclear accidents are catastrophic. The answer to that is that, while car accidents occur often, catastrophic nuclear accidents have not occurred. Moreover, because an accident might conceivably become catastrophic, we have multiple safeguards. In addition to studying the emergency core cooling system, we prevent the situation in which the emergency system would have to go into action. Even if radioactivity should actually escape from the reactor, we confine it within the housing that surrounds the reactor. Study of the emergency core cooling system has the limited purpose of increasing the reliability of one among several safety devices. If any one of the several safety devices works, trouble will be avoided.

We have mentioned that a few years ago there was appropriate criticism that nuclear development and nuclear safety were ultimately the responsibility of a single individual. In January 1975, however, safety and licensing were entrusted to an independent entity, the Nuclear Regulatory Commission (NRC). Within the first few months of this agency's separate existence, a leak was discovered in the water cooling system of one of the boiling-water reactors. Because such a leak might eventually develop into a rupture, requiring that the emergency core cooling system should take over, the small leak was taken very seriously. Not only was the reactor shut down, but also orders were given that within the next month all similar boiling-water reactors, numbering approximately two dozen, should be shut down and inspected. Within weeks this order had been carried out, at considerable cost. None of the other boiling-water reactors had any leaks.

Such vigilance should reassure the public. Paradoxically, in some cases it had the opposite effect. The simple circumstance that a great number of reactors were ordered shut down was used by some people to "prove" that reactors are unsafe.

One very real hazard, especially in California, is from earthquakes. To the original suggestion of the Advisory Committee on Reactor Safeguards, that reactors shut down automatically at the first sign of a tremor, more precautions have been added over the years. Under present regulations no reactor can be built near an "active" earthquake fault. The definition of an active earthquake fault is remarkable. To deem it otherwise proof is required that the fault has not been active in the past twenty thousand years. Such proof can, of course, never be given with complete confidence, yet it is useful to make the attempt.

We should also realize that reactors have a much greater inherent resistance to earthquakes than most man-made structures. Human habitations, of course, are hollow. They are apt to collapse in a severe earthquake. A reactor, by contrast, has essentially no cavities. An earthquake will shake the reactor as a whole, and is not apt to damage it. Probably the only big man-made structures that are safer than nuclear reactors are Egypt's great pyramids, which have stood for six thousand years.

A particularly vulnerable structure in an earthquake is a dam, including dams constructed for hydroelectric purposes. Water from collapsed dams has killed thousands of people. A catastrophe of this kind was narrowly avoided on February 9, 1971, in the San Fernando, California, earthquake. The two dams in the San Fernando Valley were so badly damaged that they had to be dismantled. While the water was drained, eighty thousand people in a twenty-square-mile area were evacuated for two days because no one knew whether the dams would hold. It is believed that the dams did not promptly collapse because at the time of the earthquake the water level was relatively low.

The recently built earth-filled Teton Dam collapsed in Idaho on June 5, 1976, just three days before the decisive vote on California's antinuclear Proposition 15. Property damage exceeded a billion dollars, while only sixty million dollars had been spent on the dam's construction. Fortunately the dam broke in the daytime and there was ample warning; only ten persons were killed. At night the consequences would have been far more tragic.[5]

We have had lots of experience with dams, but relatively little scientific evaluation. In the case of nuclear reactors the opposite is true: we have had limited, though still substantial, experience, but extremely careful scientific evaluation. It is likely that reactors are among the few structures that would withstand a severe earthquake and would be promptly available for further service.

This argument should not, of course, be used to indiscriminately forbid dam construction. Besides producing electricity, dams prevent natural

[5] The Toccoa, Georgia, dam disaster in November 1977 claimed thirty-eight lives.

floods. Here, as in all cases, dangers and benefits should be balanced, and safety measures improved.

There is hope for even greater safety in the future. Indicators of impending earthquakes are being studied and developed.[6] In early February 1975, the Chinese used such indicators to take precautionary measures in a populous major industrial area of southern Manchuria.[7] The earthquake there registered 7.3 on the Richter scale, which measures tremors and quakes on a scale where each additional unit means a factor ten in energy; Richter 7 has ten times the energy of Richter 6. Yet extensive casualties were prevented because the populace had been prepared or evacuated in advance. Before a predicted earthquake, a reactor would be shut down in advance of the time that the earthquake is expected. After the passage of several hours in such a shut-down condition, a reactor would be completely safe. Loss of electrical energy production from reactors during the dangerous period should be gladly accepted.

Accidents constitute the only real danger from reactors, and by use of careful and well-thought-out procedures this danger has been reduced to an immeasurably small probability.

In public discussions, however, a threat of danger from "nuclear ashes" figures almost as importantly as the danger of a nuclear accident. It is claimed, for instance, that radioactivity from one element, plutonium, will continue to be a hazard to our descendants for more than one thousand human generations. Human institutions change, and some believe that within thousands of years a situation might develop whereby the plutonium could poison a vast number of human beings.

The lifetime of plutonium is indeed 24,000 years. This element is more valuable than gold, however, because it can serve as a nuclear fuel. Instead of constituting a menace for a thousand human generations, it should be burned up in a nuclear reactor in not more than one generation. But we need to understand what actually is going to happen to spent fuel elements, the so-called "nuclear ashes." Because of their re-

[6]One indicator depends on a decrease in sound velocity for compressional waves in the earth's crust due to the opening of cracks. Later the sound velocity increases as water fills these cracks. Then the earthquake is apt to follow in a day or two. An indication used in China is the rapid lowering of water levels in wells, which occurs when water enters the cracks.

[7]A more recent earthquake in China unfortunately was *not* predicted. Many people perished. Earthquake prediction is not yet an exact science.

sidual radioactivity, spent fuel elements continue to produce heat as they leave a reactor. They are kept under water, which serves as a coolant and also prevents the escape of radiation. This step in itself would appear to be a satisfactory solution, except that spent fuel elements accumulate year after year, and will, in time, become too numerous to contain in temporary storage. In a couple of decades, if not earlier, we should have permanent storage for radioactive wastes.

There is a popular misconception about the existence of adequate means of disposing of spent fuel elements. In reality the reason for continued anxiety is that no definitive method has as yet been *adopted* for the treatment of these materials. To achieve permanent disposal, a lot of time and care must be spent. For this reason, and probably for reasons of bureaucratic delay and confusion as well, details of the disposal procedure have not yet been completed. Some observers, therefore, have the impression that no satisfactory methods of final disposal exist. The fact is that several inexpensive methods are under careful consideration. Their cost is not likely to add more than one percent to the expense of generating electricity. The only difficulty remaining is to make a choice and work out details.[8]

I want to describe the proposed storage method which to me sounds best. After spent fuel rods have cooled for a year or two they produce considerably less heat. Then they can be dissolved and the heavy, valuable, relatively long-lived elements can be separated from the light ones. The heavy elements consist of remnants of the original nuclear fuel and other elements obtained by the capture of one or more neutrons in the original fuel. To this must be added their radioactive decay products, among which plutonium 239 is by far the most important.[9]

---

[8] The difficulties of making a choice were discussed in approximately A.D. 1200, by the French philosopher Buridan. His teachings are best remembered in the story of Buridan's ass. The poor beast starved to death between two equidistant and equally desirable stacks of hay. In the case of spent nuclear fuel elements we have more than two choices. We also have four asses, popularly known as DOE, NRC, EPA, and CEQ, which furthermore, are yoked together and must act in unison. A further complication is that the present administration is taking a position against the "reprocessing" necessary to separate the plutonium. The reason is fear that plutonium could be used to make nuclear explosives. Actually, nuclear explosives could be produced in many other ways, while lack of reprocessing causes difficulties in the nuclear fuel supply and will perpetuate the fear of nuclear ashes.

[9] Uranium 238, the main portion of uranium, captures a neutron and gives uranium 239. Two beta decays follow, giving plutonium 239. This phenomenon is discussed later in this chapter.

Plutonium can hardly be called nuclear ash. It is "flammable," and even in the reactor some of the plutonium is burned up before the fuel is removed. Many of the heavy elements are highly useful. Most of them have long lifetimes, and in the course of time all of them are likely to find important employment. They will be carefully guarded, not only because they are dangerous, but also because they are valuable.

Usable fission products will also have to be separated. These fission products, the fragments into which the uranium has split, are the real nuclear ashes. While many of these fragments are usable, they are produced in such abundance that some of them, possibly the major portion, will have to be stored. Because of their radioactivity they still produce some energy, but this energy output is so small that it is not worth utilizing. (Uses of fission products will be discussed shortly). After separation of the heavy elements and usable fission products, the remainder can be incorporated into a practically insoluble borosilicate glass.[10] This process has been developed and we know that such glass can be made to consist of approximately 30 percent fission products and 70 percent glassy substance, mostly silicon, boron, and oxygen. As to volume, the material can be handled without much difficulty. Assuming a full nuclear economy in the United States, it has been estimated that the amount of this glass that would be produced in the last quarter of the twentieth century would cover a football field to a depth of ten meters.

I think it would be best to place this material a mile underground in a dry and stable—that is, earthquake-free—geological layer. It could be placed in cavities within salt deposits, for instance. It is quite easy to wash out such cavities. Salt has the added advantage of affording protection from water. If, in spite of the original geological evaluation, an earthquake should occur, any rift in the salt is apt to heal. Thus, we would achieve triple protection. Water cannot get in through the salt, but if

---

[10] Glass would not be insoluble in the "hydrothermal" solutions that could arise from the combination of radioactive decay heat and water encroachment into a closed repository. See G. J. McCarthy et al., "Interactions Between Nuclear Waste and Surrounding Rock," *Nature*, Vol. 273, No. 5659 (1978).

The heat output of the glass can be lowered, of course, by dilution. Also, advanced mineral-like waste forms that can survive under even these severe conditions are being developed at Pennsylvania State University by McCarthy and his associates. Gregory J. McCarthy, William B. White, and Diane E. Pfoertsch, "Synthesis of Nuclear Waste Monazites, Ideal Actinide Hosts for Geologic Disposal," *Materials Research Bulletin* (University Park, Pa.), Vol. 13 (1978).

water should, somehow, reach the material, the radioactivity would be in an insoluble form and, most importantly, the material would be a mile underground. Water, the most plausible means of transporting radioactivity, would not be available to the radioactive substance.

After approximately three hundred years in this deep burial place the radioactivity would have become less than that of the original uranium ore from which it was derived. It would continue to decay through passing time. The crust of the earth would actually be depleted of radioactivity. Thus, we would have permanently freed the biosphere, the region of the living, from the radioactive products.

Uses of fission products, the real nuclear ashes, are many, both actual and potential. As tracers they follow the complicated paths of similar atoms in a variety of processes, including biological processes. They can be used in the preservation of foodstuffs, without radioactivity being transferred to them. Numerous industrial uses have been developed, as in measuring and controlling the thickness of some mass-produced foils, or in observing the erosion of bearings in machinery. In many of these applications, the essential property of the fission products is that they are observable in extraordinarily small amounts. Safe handling of radioactivity, in fact, is possible because one-millionth of what is harmful can be readily observed. Peculiarly enough, what should serve as reassurance manages to scare people. Few objects seem to be as frightening as a clicking Geiger counter, even though an older type luminous-dial watch would set it to clicking with great vigor.

Let us consider some other industrial uses of fission products. There are varied examples illustrating the wide scope of possible applications. Fission products are now used in hundreds of ways. In the future, thousands of applications are to be expected.

We can illustrate the use of radioactive fission products as tracers in an improbable sounding case. When your clothes go the the cleaner, spots are not actually removed, at least not completely. They are more evenly distributed over the garment or perhaps shared by the clothes of other customers. This happens not because the people at the laundry are malevolent or negligent, but because chemistry is an imperfect art. A good detergent will minimize the spot distribution, and should retain the dirt in the cleaning solution. But how do detergent manufacturers find out how well they accomplish this purpose?

What can be done is to add a small amount of radioactive substance to the material that makes the spot, so it will share the fate of the spot. It can then be determined in a quick, quantitative manner what fraction of the spot and the radioactivity remains attached to the clothes, and what fraction is washed away. Do not be nervous about this; no radioactivity will be attached to your spot. This procedure is used only in tests.

For Appalachia, a very different use of fission products has been proposed. The forests in that region produce soft wood that is all but valueless. Radioactivity can be used to change this soft wood into hard wood suitable for floors or furniture making. The soft wood is soaked in an organic solution that penetrates between its easily separated fibers. Before the soaking a short-lived radioactive substance is added in moderate amounts to the solution. The effect of radioactivity on the organic solution is to "polymerize" the molecules, to join them into long chains. The polymerized substance will tie wood fibers together. These wood fibers can no longer be separated easily; soft wood will have been effectively turned into hard wood. Meanwhile the radioactivity will have decayed to the vanishing point. If the half-life of the radioactivity used is one day, for instance, in a month only one-billionth of it will remain. We should note that the use of short-lived fission products, while very interesting and important, will not contribute to the disposal problem. Whether a short-lived radioactive substance is employed to cure thyroid cancer or used on wood in Appalachia, it will have to be produced for that special purpose. Extracting it from spent fuel elements would be impractical because it would vanish before it could be utilized.

An extremely important use is to tie radioactive substances to chemicals that seek cancerous cells in the human body even after metastasis has occurred, that is, after cancer has spread from its first location to some other part of the body. This has been done successfully with iodine in the treatment of cancer of the thyroid.

My friend Lowell Wood suggested exploring a similar possibility. Though only a little work has been done on it, it could be extremely important. There is some evidence that the widely used antibiotic tetracycline tends to attach itself to cancerous cells. This seems to be the case in different kinds of cancer. Radioactive tetracycline has actually been used in experiments with rats. Lawrence Livermore Laboratory in California has a good supply of ultraheavy hydrogen, tritium. It also has good methods of handling the radioactive tritium. In experiments on rats with spontaneous breast tumors, untreated animals died, while 70 percent of the treated animals recovered. The experiments were discontinued, which I regret. Some time they should be resumed. I am not sure that tetracycline is "it." But it would be astounding if some more or less complicated molecule is not found that tends to attach itself to cancer cells and thus serve as an effective carrier of radioactivity.

As a final example I should mention caesium, an abundant radioactive fission product with the rather long average lifetime of a little over forty years. Caesium emits penetrating gamma rays. An important and extensive proposed use is to irradiate sludge, particularly sludge with a high organic content.

Sludge as a rule is not only useless, it is also dangerous because of the many bacteria and other microorganisms existing and breeding in it. Irradiation by the gamma rays from caesium will kill all the microorganisms without essentially altering the sludge and, of course, without rendering it radioactive. The radioactive caesium is not mixed in with the sludge; it is in solid containers, probably in the form of needles. Gamma rays not only manage to escape from the needles, but penetrate deeply into the sludge. As a result, the sludge is sterilized and thus not only becomes harmless, but may in some cases be used as animal fodder.

In spite of the many uses of radioactivity, it is virtually certain that a considerable amount will remain that must have safe disposal. Recovery of this radioactivity is not particularly desirable because the nuclear industry is apt to produce enough fission products for new uses as they are discovered.

If nuclear accidents can be prevented and radioactive ashes buried, what more is there to worry about? Opponents of nuclear reactors are never at a loss; they will tell you that low-level radiation could kill thousands, maybe millions of people.

That statement sounds, and is, extravagant, yet millions believe it. What are the facts? The most significant fact is that we know more about radiation hazard than about any other contemporary hazard.[11] Chemical poisons are specific. But a small alteration in the structure of a molecule can change it from a food or curative drug into a poison or a carcinogen whose effect may be delayed, and consequently hard to verify. Chemicals act on biological systems the way a key acts on a lock. A small variation in the key might destroy the desired effect or may cause it to stick in the lock, thereby blocking the desired function, which in our case is a needed

---

[11] We continue to seek comparable knowledge about food additives, medications, and pollution from coal or petroleum, for example.

chain of biological events. Our knowledge of biology does not suffice for predicting how a molecule will affect the human body.[12]

The effects of radiation are much simpler. Radiation acts like a hammer; it breaks molecules. And different kinds of radiation—gamma rays, beta rays, or alpha particles—act quite similarly. The energy deposited by radiation in a given volume of specific tissue is a good measure of the damage to be expected. Furthermore, it is relatively easy to find out what amounts of radiation will get to any tissue, carried by a specific radioactive substance. Gamma rays are the most important among external sources, since they penetrate deep into the body.[13]

If the exposure were substantial, death or radiation sickness would result. The consequences are well understood, and high radiation doses must of course be avoided. When radiation doses fall to a low level they can still cause mutations, thereby damaging offspring. They may also contribute to cancer after some delay, which, as a rule, is a number of years. Even though these effects are small, and therefore difficult to observe, we know a lot about them. The reason for that is simple: experience. All of us are exposed to background radiation from natural sources, including cosmic rays, radioactive potassium in our blood, or radioactive traces in natural food that are derived from uranium in the soil. Most of us get additional slight radiation from diagnostic x-rays.

Radiation is commonly measured in *roentgens* or *r-units* (sometimes called "rems"). The radiation that an American receives from nonnuclear sources amounts to 0.17 r-units per year. If one thousand r-units were received over the whole body, death would surely result. Half that dose would give a 50 percent mortality rate. Below that dose the danger of death decreases rapidly but radiation sickness—the symptoms of which are changes in blood composition and damage to the intestinal walls, accompanied by vomiting and a loss of hair—is observed at a dose as low as one hundred r-units. Below that amount no immediate biological effects are observed, but genetic changes and delayed development of cancer will be present down to an undetermined lower level.

The average dose of 0.17 r-units per year to which we all are exposed from natural background and diagnostic x-rays does not produce any

---

[12]The situation is even more complicated in the case of allergens that act differently on different people. Biological hazards appear to be quite unpredictable.

[13]Neutrons are equally important, but exposure to neutrons is not expected as a consequence of the operation of nuclear reactors.

MOTHERS LOOK OUT FOR YOUR CHILDREN!
ARTISANS, MECHANICS, CITIZENS!
When you leave your family in health must you be hurried home to mourn a
DREADFUL CASUALITY!
PHILADELPHIANS, your RIGHTS are being invaded! regardless of your interests, or the LIVES OF YOUR LITTLE ONES. THE CAMDEN AND AMBOY, with the assistance of other companies without a Charter, and in VIOLATION OF LAW, as decreed by your Courts, are laying a
LOCOMOTIVE RAIL ROAD!
Through your most Beautiful Streets, to the RUIN of your TRADE, annihilation of your RIGHTS, and regard less of your PROSPERITY and COMFORT Will you permit this? or do you consent to be a
SUBURB OF NEW YORK!!
Rails are now being laid on BROAD STREET to CONNECT the TRENTON RAIL ROAD with the WILMINGTON and BALTIMORE ROAD, under the pretence of constructing a City Passenger Railway from the Navy Yard to Fairmount!!! This is done under the auspices of the CAMDEN AND AMBOY MONOPOLY!
RALLY PEOPLE in the Majesty of your Strength and forbid THIS
OUTRAGE!

(a)

WHAT DO YOU DO
IN CASE OF A
NUCLEAR ACCIDENT

KISS
YOUR CHILDREN
GOODBYE

(b)

FIGURE 10-2 *(a) Exploiting fear of the unknown is not a new technique for opposing technical innovation. Emotional scare protests are part of the history of America's railroads, as this poster shows. (b) A modern technique for exploiting the fears of the uninformed is understated, for even greater effect.*

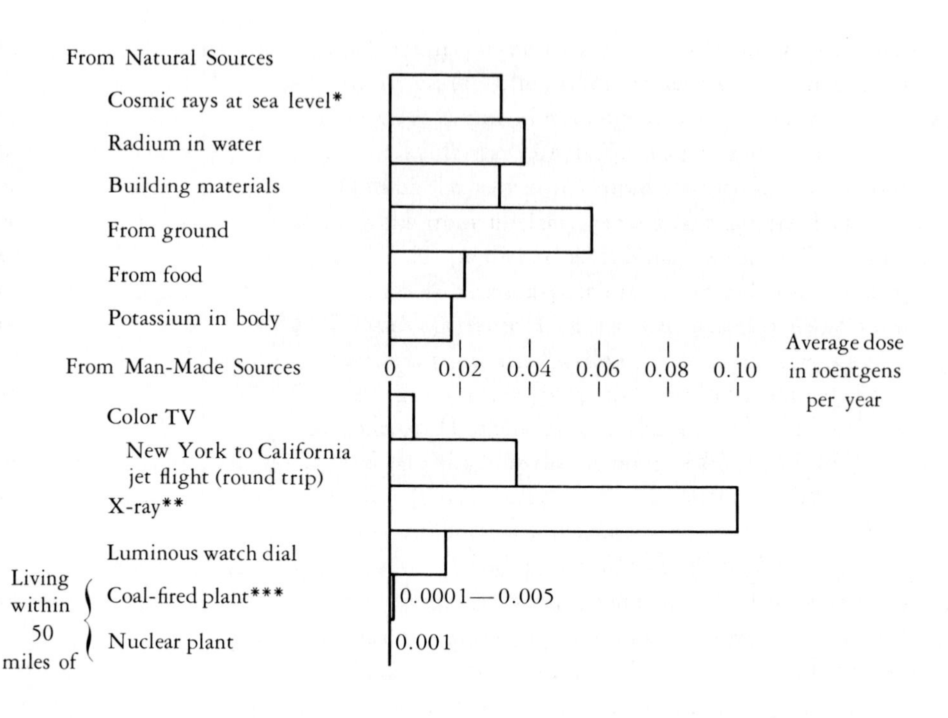

FIGURE 10-3 *How people get radiation.*

noticeable effects. Beyond a slight increase in mutations and a possible contribution to "natural" cancer incidence, there may actually be no effects. Experiments on rats and mice have shown beneficial effects at levels much higher than the natural background; these beneficial effects are not understood and may not be applicable to humans. They show, however, that we know next to nothing about these exceedingly low-level radiations except the obvious and important point that their effect is very small.

Here is the relevant point concerning nuclear reactors: a bystander is not apt to receive radiation from a reactor amounting to more than 1 percent of natural background radiation, that is, 1 percent of 0.17, or 0.0017 r-units. On the average he will get less than that amount. The effects of natural backgrounds are so small that they remain unknown, unverified by explicit experiments in spite of diligent efforts to determine them. It is this unknown source of possible trouble that is played up as an important effect. The unknown seems to be frightening even when we know that it is much less dangerous than what nature itself provides. This is a remarkable feat of scare propaganda.

One of the very few people who make such extravagant statements about risks from nuclear reactors is a double doctor—in physics and in medicine. He predicts that nuclear reactors will produce twenty thousand added cancer cases per year in the United States and up to a million genetic deaths per year. He arrives at these figures in a unique way. He uses a "maximum permissible dose," determined in such a way that none of our tissues can receive more from artificial sources than they get from natural sources. He then assumes that in a nuclear economy all people will be exposed to this "maximum permissible dose" (which would double the radiation all of us receive). He also assumes that all effects of weak radiation are proportional to the intensity of the radiation. Adding a pessimistic evaluation of all effects, the resulting huge figures are derived.

The most obvious mistake in this reasoning is to assume that the "maximum permissible dose" will be reached. According to present or foreseeable practice, the population is and will be exposed to less than one-thousandth of this dose from the function of reactors. This mistake, a thousandfold overestimate, the double doctor committed with full knowledge of the facts.

There is another illustration of how to evaluate such propaganda efforts. At a hearing on the safety of a reactor in Illinois, Dresden III, a vociferous opponent raised the issue of weak radiation. A young representative of the Atomic Energy Commission asked him, "Can you tell me which would give you more radiation, to lean against Dresden III for a full year, or to sleep with your wife each night?" The opponent seemed confused. The man from the AEC continued, "I don't want to imply that your wife is particularly dangerous, but she has radioactive potassium in her blood. Of course, you get additional radiation from your own blood, and hers will get to you only after the radiation penetrates some shielding.[14] Dresden III has much more radioactivity than your wife, but it also has much more shielding. I just want to know which would give you more radiation." There was still no reply.

---

[14]Beware of lean wives! (This is not my slogan, but should be taken up by nuclear scare-mongers.)

Shortly after the hearing I received a brief report from the AEC. "The comparison was made. You get more radiation from Dresden III than from your wife. Therefore I shall not recommend a law requiring that married couples sleep in twin beds. But I should warn against the habit of sleeping each night with two girls, for then you would get more radiation than you get from Dresden III."

The discussion here on weak radiation may seem too long. Its length corresponds to the public emphasis upon this topic. Objectively we could have dismissed it in two short sentences: "There is no reason to worry about weak radiation from nuclear reactors. That radiation is negligible compared to the natural background radiation that has acted on living things as long as they have existed."

Having tried every possible argument to show that nuclear reactors are dangerous, opponents also want to prove that nuclear reactors are useless. In particular, they ask if there will be enough nuclear fuel to run them. At present, reactors utilize essentially only one rare isotope, uranium 235, amounting to less than 1 percent of uranium. We have learned economical ways to mine ores containing 0.16 percent uranium. What burns up in the reactors is about one part in one hundred thousand of what is mined. If we fail to improve these procedures, our nuclear fuel may not last very much longer than our petroleum reserves. The price of uranium ore has already risen. In oil equivalents, it was 25¢ per barrel of oil not many years ago. Now it is between $1.50 and $2.00 per barrel. Does this indicate an unavoidable trend?

Opposition to nuclear reactors by a vocal minority has something to do with the developing shortage. Uranium exploration pays off only after the passage of years. If there is a possibility that nuclear reactors may be banned, then why continue to search for more uranium?

In 1945 ore had to contain 2 percent uranium to be exploited. Today, after technical improvements in mining procedures, 0.16 percent suffices. We have learned a lot about utilizing relatively poor ores. And there is lots more poor ore than rich ore, so that the total uranium in poor ores amounts to a huge reserve. If we could utilize ore with even lower uranium content there would be enough uranium. But this means more research, and why invest in research when opponents of nuclear power may win?

These same opponents use their prediction of a uranium shortage as an argument against nuclear power. It is an excellent example of a prophecy that may become self-fulfilling.

In fact, however, nuclear fuel is going to suffice even without reliance on poorer ore, if only all of the uranium can be used—the abundant (99.3 percent) uranium 238 as well as the rare (0.7 percent) uranium 235. Plans to use uranium 238 go back to World War II, when "breeder" reactors were first conceived. Progress on these breeder reactors has been disappointingly slow, however. Year after year an economically successful breeder moves farther into the future. Today the optimists hope it can be a reality in fifteen years. Never before has success seemed so far away.

What is the reason for this? And what is a breeder reactor?

At first glance it appears that the abundant isotope uranium 238 is less than useless in a nuclear reactor. It not only fails to deliver the considerable energy characteristic of fission, but it also absorbs the neutrons essential to a chain reaction. When uranium 238 absorbs a neutron and turns into heavier uranium 239, however, we have an unstable radioactive nucleus. It has too many neutrons, and relatively too few protons.[15] The neutron-absorption process is followed by two successive radioactive decays, and yields the highly fissionable plutonium 239. Comparable in its properties to uranium 235, plutonium 239 can be used in fission reactors.

The next question is, do we get enough plutonium? If we burn up one uranium 235 or one plutonium 239, do we produce just one plutonium 239, or more, or less? The answer is that it depends on the speed of the neutrons in the reactor. If, before the neutrons are captured, they are slowed down by collisions with other nuclei, we get fewer plutoniums. If we operate with fast neutrons, we may get 1.3 or even 1.4 plutoniums for every plutonium burned up. This is the function of a *fast* breeder, which works with fast neutrons.

Fast neutrons unfortunately are stubborn; they are not easily influenced. They have approximately the same behavior no matter what material they encounter. Furthermore, they produce fission only after they have traversed a long path in plutonium. The result is that great amounts of plutonium are needed in a single reactor.

Slow neutrons have been slowed down by repeated collisions with light nuclei, from an original speed of one-thirtieth of the velocity of light to a much lower velocity, comparable to that of sound. They behave very differently from fast neutrons. In some elements, such as cadmium or boron, they are easily absorbed. Therefore, these atoms can be used in

[15] Uranium 239 has 92 protons and 147 neutrons.

control rods that regulate the activity of a reactor. In fissionable materials such as uranium 235 they cause fission with high probability. Relatively little uranium 235 suffices to build a reactor using slow neutrons. For a fast reactor much bigger quantities are needed, particularly if vigorous breeding is wanted, for example, replacing one plutonium by 1.3 plutoniums in the breeding cycle. That rate is possible only if few neutrons are lost through outside surfaces. A good fast breeder needs two tons of plutonium—an impressive amount.

Since we have no really good absorbers for fast neutrons, many further engineering difficulties arise. Moreover, we must not slow down the neutrons, and this constraint rules out the use of elements of low atomic weights; even in an elastic collision they rob the neutrons of a considerable fraction of their energy. Hydrogen, of course, must particularly be avoided.

In 1945, when plans were laid by Enrico Fermi, Eugene Wigner, and others, for the long-term development of fission reactors, the fast breeder seemed the obvious answer. Total available uranium resources then appeared quite limited. Only breeding, utilizing most of the uranium rather than only the rare uranium 235, seemed to hold promise; and the high breeding ratio of fast breeders seemed especially attractive. Since that time uranium resources have greatly increased, due partly to new discoveries and partly to the exploitation of poorer ores. Therefore, we can no longer say that the fast breeder is necessarily the right answer.

We have excellent reasons for modifying the conclusions reached by the most outstanding people in 1945. One is that, though we have pursued the fast breeder program for more than three decades, an economically competitive fast breeder is still not more than a promise for the future. Another reason is that opponents of nuclear reactors are most worried about fast breeders. They say, with some justification, that work on them requires lots of money for research and development. Also, they are frightened at the prospect of a "plutonium economy," a term of unclear meaning that is part of the vocabulary of "antinukes."

The central issue is whether we are going to suffer from nuclear fuel shortages, as the antinuclear movement asserts. If the fast breeder should work, the fuel problem would be solved; and it is quite possible that the fast breeder will work. There is an alternative, however, which is almost certain to work. The statement that sooner or later the nuclear industry will have to turn to fast breeders is simply not true.

Instead of relying on the uranium-plutonium cycle, we can use the thorium-uranium cycle, which starts with the abundant element thorium. The two cycles are analogous; instead of uranium 238 → uranium 239 → neptunium 239 → plutonium 239, we use thorium 232 → thorium 233 → protactinium 233 → uranium 233. We start with thorium 232, which is found in nature.[16] Neutron absorption yields the heavier, unstable thorium 233, which decays first into protactinium 233 and then into uranium 233. Uranium 233 is similar to plutonium 239; both can be used in reactors and both produce fission with neutrons of any energy. They differ, however, in two small characteristics that have important practical consequences.

One difference is that plutonium emits more neutrons, resulting in a better breeding ratio. Indeed, this was why plutonium was preferred in 1945.

The second difference is in the behavior of thorium and plutonium when they are subjected to slow neutrons. Thorium functions more efficiently. Plutonium does much less well because it not only gives fission, but also absorbs a considerable fraction of the neutrons uselessly. For plutonium, fast breeders become a necessity.

Thorium, which Wigner advocated from the beginning, serves with little difference whether in fast breeders or slow breeders. The Canadians have developed the "Candu Reactor," almost a breeder, which uses heavy water to slow down the neutrons.[17] Starting from one uranium 233, it can give just one uranium 233. Of course, uranium 233 is not found in nature. But the reactor can first be charged with uranium 235 and then switched over to the thorium cycle; neutrons are absorbed in thorium, give uranium 233, and from then on, uranium never need be fed into the reactor. The reactor can run indefinitely on thorium.

The thorium cycle is a practical and economical possibility today—at least in the form of a realistic blueprint. We certainly need not run out of nuclear fuel, at least not before the next ice age.[18] Scarcity of fuel is as

---

[16]Thorium has only one stable isotope. It is an element that occurs in many minerals.

[17]The unavailability of heavy water in big quantities is the only obstacle to widespread use of this reactor. There has been little recent research into the production of inexpensive heavy water, but research is really hopeful.

[18]There are two other reactors which could use the thorium cycle in the near future: the high-temperature gas-cooled reactor (HTGR) and the epithermal reactor. Like Candu, these exploit neutrons effectively. They also have the advantage that they do not suffer from the NIH (Not Invented Here) disease.

weak an argument against reactors as the other objections we have discussed.

Two serious questions remain. Both concern the misuse of nuclear energy.

In December 1973, in a provocative series of articles in *The New Yorker*, John McPhee discussed possible acts of nuclear sabotage, and the possibility that plutonium produced by nuclear reactors might be stolen and used to construct a homemade nuclear explosive. With the continuing prevalence of terrorism, these dangers should not be disregarded.[19]

Terrorists do not necessarily want to kill the maximum number of people that they can. They seek to accomplish their aims through acts of terror. Nuclear weapons or even homemade nuclear explosives may not be as dangerous as means more available to terrorists. But people have been conditioned to fear the nuclear menace more than any other single danger.

McPhee's *New Yorker* articles were, in fact, justified. At the time they were written nuclear reactors and their products, which could be used in bombs, were guarded in a manner less than completely satisfactory. Considerable change has occurred since. Those articles shook up the authorities. Vigilance was redoubled and refined. Today the danger has decreased; it is no longer significant.

It has been argued that reactors and nuclear materials cannot be guarded adequately without turning our country into a police state. But compare this worry with the problem of airplane hijacking. Skyjacking is incomparably easier to perform than acts of nuclear terrorism would be. Defense against skyjacking is made particularly difficult by the need to screen many millions of passengers. With respect to security, as the number of essential persons decreases, safeguards become easier to introduce, more feasible, and more effective. There is no question that reactor products can be guarded. This difficult task, which a few years ago was performed somewhat perfunctorily, is now carried out in a satisfactory manner.

I wish I could be as optimistic and positive about the remaining objection: as nuclear reactors spread among nations their production will

---

[19]Such possibilities were discussed during the first meeting of our Reactor Safeguard Committee in England. John Wheeler remarked, "Maybe a saboteur is present right in this room." Klaus Fuchs, a member of the British delegation, was present and listening to this remark. A few weeks later he was arrested. He turned out to be Russia's most important spy.

enable almost every country to acquire nuclear weapons. This statement, most unfortunately, is true. I believe that eventually nuclear proliferation is unavoidable unless we find far better solutions to international problems than are now on the horizon. This is the main worry influencing our present administration.

Should the United States therefore abandon this dangerous business? It is my firm belief that if we dropped out of nuclear energy development it would have only a minor effect on slowing down the proliferation of nuclear reactors. Today only one-third of the nuclear power reactors are to be found in the United States, which has about seventy in operation; and rapid expansion continues in other countries. If we ceased participating we would no longer have the influence we now use to stabilize the world situation.

Preventing war is of the utmost importance. Most particularly we want to avoid nuclear war. United States policy for at least two decades has been to decrease the danger of nuclear conflict by encouraging various arms limitations. This policy has been unsuccessful. Its chief result has been that we have lost our leadership in the military field to the Russians. This can hardly be called an advantage. Peace may well depend on the power held by those who desire peace.

There is another way in which peace can be reinforced. That is to diminish the reasons for conflict. This goal has been pursued by the United States through foreign aid and other means. It cannot be argued that this path is easy. The great advantage of this course, however, is that practically all of its effects are positive.

Shortage of energy is a painful problem that now affects the great majority of mankind. The United States suffers from it, Western Europe is affected seriously; in Japan the shortage is even greater. But those most terribly hit are the underdeveloped nations. For them, energy shortage translates into shortage of food. Effective agriculture is impossible without fertilizers, without water, and without mechanized equipment. All of these require energy.

We are deeply conscious today of the dangers of pollution. Of all pollution, of all defilement, the most widespread and the worst is pollution by poverty. Three-fourths of mankind is suffering from it. If the energy shortage continues there will be no cure; if nuclear reactors are abandoned, the energy shortage will increase.

There is in fact no prospect that nuclear reactors will be abandoned throughout the world. If the United States continues to participate in

TABLE 10-1 *Nuclear Power Reactors Worldwide*

Country	*Operating*	*Under Construction*	*Ordered*	*Planned*	*Total*
Argentina	1	1	3	0	5
Austria	0	1	0	1	2
Belgium	3	2	2	0	7
Brazil	0	3	0	6	9
Bulgaria	2	1	1	4	8
Canada	7	10	4	5	26
China (Taiwan)	0	4	2	0	6
Czechoslovakia	1	4	0	16	21
Denmark	0	0	0	6	6
Egypt	0	0	0	5	5
Finland	0	4	0	0	4
France	10	17	12	8	47
Germany (West)	3	0	2	0	5
Germany (East)	7	12	8	4	31
Hong Kong	0	0	0	1	1
Hungary	0	1	1	0	2
India	3	5	0	0	8
Indonesia	0	0	0	3	3
Iran	0	0	4	1	5
Ireland	0	0	0	1	1
Israel	0	0	0	1	1
Italy	3	2	4	16	25
Japan	10	14	0	5	29
Korea (South)	0	1	1	8	10
Luxembourg	0	0	1	0	1

development of nuclear reactors, in production of nuclear fuels, and in reasonable use of nuclear "wastes," there is greater hope that the worldwide problem of possible misuse of nuclear energy will be solved.

In all other respects opponents of nuclear power in the United States have no valid argument. In the case of nuclear proliferation the argument is real and frightening, but the answer proposed by antinuclear propagandists is the answer of the ostrich: "Let's hide our heads in the sand."

TABLE 10-1 *Continued*

Country	*Operating*	*Under Construction*	*Ordered*	*Planned*	*Total*
Mexico	0	2	0	7	9
Netherlands	2	0	0	3	5
Pakistan	1	0	0	1	2
Philippines	0	0	2	8	10
Poland	0	0	0	2	2
Portugal	0	0	0	4	4
Rumania	0	0	1	2	3
South Africa	0	0	2	0	2
Spain	3	7	7	21	38
Sweden	5	6	0	3	14
Switzerland	3	1	3	2	9
Thailand	0	0	0	3	3
Turkey	0	0	0	1	1
USSR	19	8	0	10	37
United Kingdom	29	10	0	7	46
Yugoslavia	0	1	0	1	2
Cuba*	0	0	0	8	8
Kuwait*	0	0	0	2	2
Libya*	0	0	0	2	2
New Caledonia*	0	0	0	2	2
45 Countries	112	117	60	180	469
USA (1978)	71	83	61	1	216

*Details of implementation of planned programs are not available for these countries.
Source: Data from Atomic Industrial Forum.

Antinuclear forces selected California for a first major test. They tried to ban nuclear reactors under the guise of making them safe. Their Proposition 15 on the California ballot amounted to a prohibition of future nuclear reactors, and would have reduced the power of existing reactors to a level that would hardly have been more safe but certainly would have been less economical.

Why was California selected? In that state's economy nuclear reactors

FIGURE 10-4 *Serving the Sacramento Municipal Utility District, the Rancho Seco nuclear-fueled electric power plant in central California has a capacity of 913,000 kilowatts. Seemingly dwarfed by the two hyperbolic cooling towers, the reactor building (left) stands 40 feet high and weighs 422 tons. The essential portion of the reactor, earthquake-resistant, is enclosed within this smaller structure. All steam-powered plants need quantities of cooling water. In the natural draft cooling towers seen here, each 425 feet high, steam from the turbine is condensed and recirculated.*

certainly play a significant role. There were also statements (not completely free of malice) that California's climate is favorable to the growing of fruits and nuts, but especially nuts.

Nearly five hundred thousand signatures were needed to put the proposition on the ballot. They were easily obtained. By the time of the vote in June 1976, only one and a half million people voted against nuclear reactors: three million voters supported nuclear reactors by voting against the antinuclear proposition. It seems that even in California the nuts are in a minority.

At the time of the 1976 presidential elections, six more states had antinuclear propositions: Oregon, Washington, Colorado, Arizona, Montana, and Idaho. All were defeated by an average margin of two to one.

Even these voter mandates did not end the controversy. Scare propaganda continues unabated. It seems that developing good and safe technology in order to produce more energy is not all that is needed. In a

FIGURE 10-5 *"Those who are putting heavy pressure on our current administration to put the nuclear genie back in the bottle are not well informed when they offer expensive solar as the alternative."—Aden B. Meinel and Marjorie P. Meinel (pioneer workers in the solar energy field),* The Energy Daily, *August 4, 1977. (George Bing.)*

democracy it is also necessary that citizens should understand what is going on.

Achieving understanding on the part of the people may be the most difficult problem of all. The work of an educator is not unlike the task of Sysiphus. If you have answered a number of questions your next job is to answer them all over again.

PROSPECTS FOR THE MORE DISTANT FUTURE

CHAPTER 11

CONTROLLED FUSION

In which the reader will find out how to tame and control the hydrogen bomb, either by making it slow or by making it small. Unfortunately, the work is not easy and the energy obtained may not be cheap.

Too often those who know the least about a subject are the ones who talk the most about it. This seems to be the case with controlled fusion. I would like to tell you something of its history and its physics, and I believe I know the subject.

Many people know the history of controlled fusion, but far more are unfamiliar with it. Probably more people still are unacquainted with the extremely interesting physics that accompany it. This chapter contains a little more physics than the others, for advanced branches of science are deeply interrelated with the feasibility of fusion. Readers without technical training should not be disturbed. They may consider the scientific references as remote "music of the spheres."

Everybody knows that fusion is the getting together of nuclei. Medium-weight nuclei have the lowest energy. When light nuclei are put together a great deal of energy is gained. This, we believe, is how energy is produced near the center of the sun and in other stars.

The first step in thermonuclear reactions in the sun is that two protons approach each other, as mentioned in Chapter 2. During the very short time of the approach, approximately 10^{-22} seconds,[1] a positron and a neutrino are emitted and a proton is transformed into a neutron. This event is exceedingly rare. But if it happens the neutron must stick to the proton, for only in this way can energy become available to the positron and the neutrino, and a deuteron, the nucleus of heavy hydrogen is formed:

$$\text{hydrogen}_1 + \text{hydrogen}_1 \rightarrow \text{hydrogen}_2 + \text{positron} + \text{neutrino}$$

Note that two protons do not form a stable compound nucleus, but a proton and a neutron do.

After this initial step nuclear reactions occur more easily. Their end result is the formation of the particularly stable nucleus of helium, the alpha particle.

How does the sun do it? It has lots of patience, the kind that can wait for billions of years. Also, the sun makes use of gravitation. It has enough matter and enough pressure to contain a hot and dense center in which

[1]How do you get 10^{-22} seconds? Answer: Divide a second by ten thousand, then divide by a million, divide by a million again, divide by a million for the last time, and then you have it.

even one of the least probable reactions, that between two protons, will proceed.

We do not have the sun's mass available in our laboratories; nor do we have sufficient patience. So for us a more efficient reaction is necessary. We are trying to produce fusion with nuclei lighter than helium, particularly with heavy hydrogen (H_2), called deuterium. Two deuterium nuclei can react with each other quite easily, yielding either a helium 3 and a neutron or a tritium and a proton, with some further reactions following. Alternatively, one can use a deuterium and a tritium, and release a great deal of energy making helium 4 and a fourteen-million-volt neutron, a particle that appears in almost all fusion schemes. Furthermore, the deuterium-tritium reaction has a very big cross-section. These nuclei need hardly touch in order to react, so this energy release is much easier to achieve. The three most important reactions are:

$$\text{hydrogen}_2 + \text{hydrogen}_2 \rightarrow \text{helium}_3 + \text{neutron}$$
$$\text{hydrogen}_2 + \text{hydrogen}_2 \rightarrow \text{hydrogen}_3 + \text{hydrogen}_1$$
$$\text{hydrogen}_2 + \text{hydrogen}_3 \rightarrow \text{helium}_4 + \text{neutron}$$

or in a different notation:

deuterium + deuterium → helium 3 + neutron
deuterium + deuterium → tritium + hydrogen 1
deuterium + tritium → helium 4 + neutron

More than twenty-five years ago we succeeded in using fusion in an explosive energy release, first tested on November 1, 1952. And, as some readers may remember, this test was preceded by discussions nearly as explosive as the test: "It should not be done! It cannot be done!" And then it was done.

No sooner was it done than every politician and every bureaucrat descended upon us saying, "Now you must solve the problem of controlled fusion."

Some of us said, "That will be difficult." We were not believed, because at that time everybody knew that physicists always say, "It cannot be done." That was certainly the claim during the hydrogen bomb controversy. I had not been among those who said that the fusion explosion could not be accomplished. That helped my credibility a little, but not much. "You must do it now! You can have any amount of money; just go ahead and make the apparatus to control fusion."

Admiral Lewis L. Strauss, chairman of the Atomic Energy Commission at that time, was a very intelligent banker, not a physicist. He was one of the strongest supporters of the controlled fusion concept. But he had a little confidence in me, and I managed to convince him that fusion could not be controlled within a year or two. I thought we might do it in five years, although I doubted even that. We certainly should not begin to build special apparatus on the strength of somebody's theory. It seemed desirable to me to explore a number of approaches in several laboratories. The point, obvious now to everyone, was that explosive release of fusion energy has practically nothing in common with controlled fusion except the occurrence of certain nuclear reactions.

As it now stands there are two possibilities for accomplishing controlled fusion. One is to use the nuclear fuel, either deuterium or a deuterium-tritium mixture, in exceedingly great dilution. This approach uses what I call a high-pressure vacuum. It is a vacuum because a given volume contains very few particles; but it has high pressure because the few particles that are present have enormous energies. The other approach is to go to exceedingly high densities, at least a thousand times normal liquid densities. I want to make it clear why each of these approaches is apt to work, and why in both cases we might hope to achieve controlled fusion. The fusion is easy; control is hard.

Of course, we could use deuterium at its normal liquid density and release the fusion energy in an explosion. But it would be quite an explosion, measured in megatons, equivalent to millions of tons of TNT. For such an explosion you heat the material to a temperature so high that the charged nuclei can approach each other. In order to do that a fission explosion is used as an energy source. If the high temperature is produced in sufficiently big volume, then enough added energy is generated by fusion so that neighboring volumes can be heated and explosion results. But there is no control.

One way to gain control is to slow down the reaction. If the density is reduced to approximately a hundred-millionth of normal liquid density, the reaction becomes slow enough to permit appropriate control. But atoms, or the nuclei and electrons into which the atoms will dissociate at high temperature, have under those conditions a very long mean free path. This means that they can move long distances along straight lines before they collide. They will go straight to a wall and give up their energy uselessly to the wall rather than react with each other. We must therefore find some means to confine this high-temperature gas that consists of electrons and positive ions. This ionized gas, or plasma, must be confined in a bottle that does not dissipate energy.

Such a bottle exists. I call it a magnetic bottle. It has a system of magnetic lines of force around which the charged particles spiral so they hardly ever reach the wall. We can contain the plasma for the required length of time, say, one second or several seconds, so a sizable release of energy will occur. That was the first concept.

The simplest way to realize that concept would be to take a set of parallel lines of force so that we have a tube, a bundle of magnetic lines. The only trouble is that the tube is open at the ends. What do we do about that? We can make the tube very long; and some people still talk of doing that. If it were several miles long there might be a chance to make progress. But that would hardly be an easy or inexpensive way to operate.

A second possible way was the most popular in the early days. That is to bend the tube into a ring or doughnut—in technical language, into a *torus.* Then it has no ends. But it has other problems. If a magnetic field is established in the torus so that it follows big circles, and encloses the hole in the doughnut, then in the simple arrangements we can think of, the magnetic field will be stronger on the smaller circles close to the axis; there will be an inhomogenity of the magnetic field. This inhomogenity causes a drift in the charged particles, parallel to the axis of the torus, that is, parallel to a line struck through the entire doughnut. Electrons and nuclei drift in opposite directions. This at first establishes an electric field parallel to the axis and at a right angle to the magnetic field.

These crossed electric and magnetic fields cause real trouble. In the crossed fields electrons and nuclei drift in the same direction, away from

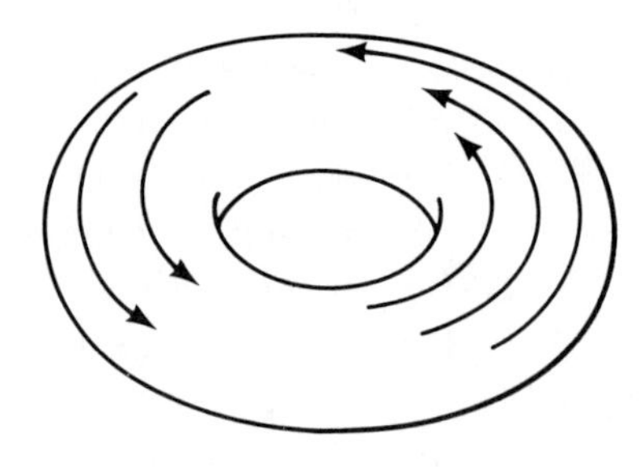

FIGURE 11-1 *Magnetic field within a torus, its course indicated by arrows.*

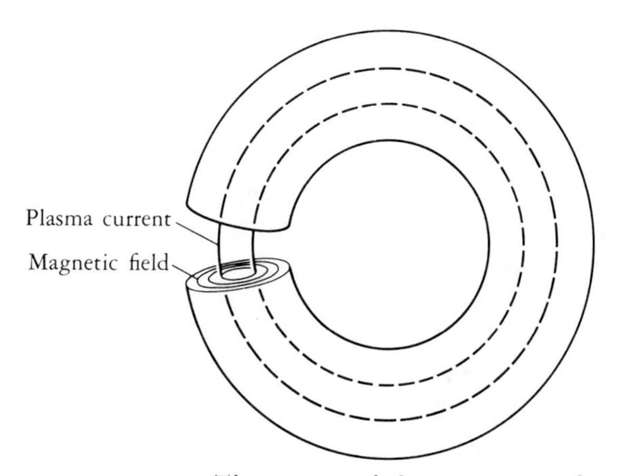

FIGURE 11-2 *The movement of plasma currents and magnetic fields, made perpendicular to each other, cause the plasma to be pinched by the tightening circles of the magnetic fields. The cutaway (left) is imaginary and serves to show the interior fields.*

the axis. Thus the whole plasma will be dumped on the outside wall of the torus.

There are various ways to overcome this difficulty. The simplest is to cause the magnetic fields not to follow the big circles, but to follow the small circles of the torus, circles that would lie on a plane within a segment if the doughnut were broken open. In this case we need a current flowing through the plasma around the torus embracing the hole of the doughnut. (Remember that currents and magnetic fields are perpendicular to each other.) The current produces magnetic fields along the smaller circles, which have a strong tendency to shrink. The plasma is carried along. The effect is that the plasma is literally pinched. The model, therefore, is called the *magnetic pinch.* The magnetic pinch has the desirable property of heating the plasma, making it denser, and creating precisely the conditions in which a thermonuclear reaction should occur.

We tried this technique and it worked practically at once. We induced the current to flow along the big circle. The plasma was pinched and neutrons were emitted, just as expected. Obviously, we had produced the desired reaction between deuterons.

We then tried to improve the neutron yield. In a themonuclear reaction, as the temperature rises, the more easily the potential barrier between nuclei is overcome, and the faster the reaction proceeds. So when you feed more current into the torus you should get many more neutrons. And we did feed more current; but we got no more neutrons.

This experimentation took place at Los Alamos, New Mexico. Interpretation of the results came from the Lawrence Livermore Laboratory in California. My friend Sterling Colgate at Livermore came up with a thoroughly unwelcome explanation: the observed neutrons were not thermonuclear. The pinch overdid its job. The magnetic fields had contracted so violently that in many places they pinched off the plasma entirely, and the continuous flow of electricity does not like to be interrupted. Very big electric fields occurred in the pinched off places, and little discharges occurred. In those little discharges the deuterons accelerated so that they produced local reactions which were not in temperature equilibrium. Under these conditions energy cannot be released on a big scale.

All of this experimentation was shrouded in secrecy. We were expected to solve the problems and produce lots more neutrons, keeping everything secret for reasons of national security. While we were doing all this in secret, the Russians were doing the very same things in similar secrecy. They got into the same troubles; they found the same explanation. But for once the Russians proved to be less secretive than we. I. V. Kurtchatov, the leader of their project, came to England and gave a talk about the things they had discovered, which we had discovered in the same way. He did not open up the whole field, but he publicized an important part of it.

From the beginning I had argued that controlled fusion would be difficult and that it should be pursued openly. I seemed unable to make any headway toward the goal of openness. Kurtchatov's talk helped. Admiral Strauss's willingness to listen to me helped. At last, in 1958, at the time of the second Atoms for Peace Conference in Geneva, the decision was made to open up the subject. I was given the very agreeable job of presenting to the Atoms for Peace Conference the sum of our knowledge gathered up to that time. The Russians also presented their findings, but they had much less than we did. They did not want to admit it, but in 1958 we were ahead of them. When Admiral Strauss, who had opposed publication, asked me a few months later if I had learned anything from the Russians, my answer was, "No."

Just over a year later we received a beautiful publication from Russia. I immediately showed it to Lewis Strauss. It was a fine piece of work, though not of great significance. It was my first proof that the Russians were really cooperating. Let me tell you what it was.

All of us knew that to obtain nuclear reactions in a low-density plasma the necessary temperature equilibrium must be maintained among most of the nuclei. But if there were equilibrium among nuclei, electrons, and radiation, then practically all the energy would go into radiation, with too little remaining in the nuclei. It is very important, therefore, to minimize the emission of radiation. That is accomplished by excluding impurities, particularly atoms of high Z values, because these radiate vigorously. (Z is the number of protons or positive charges in a nucleus.) Calculations then show that once we have only hydrogen nuclei ($Z = 1$) with the nuclei and electrons not quite in temperature equilibrium, we should expect to get the reaction going when an average energy of a few thousand electron volts per particle is reached; this corresponds to a temperature in the neighborhood of a hundred million degrees.

We also knew that one kind of radiation, *cyclotron radiation,* cannot be avoided, but fortunately it is a kind of radiation that does not matter. Electrons whizzing around the magnetic field emit extremely low-frequency cyclotron radiation. It does not easily leave the burning zone because it is so easily absorbed. No more of this radiation can ever be emitted than that corresponding to the *black-body spectrum,*[2] which corresponds to an equilibrium between the energy in particles and the energy in radiation. In the long-wavelength cyclotron radiation this black-body radiation is weak and does not cause much loss.

Russian scientists found a mistake in this argument. In order to get a fusion reaction very high temperatures are essential. At such temperatures electrons move almost relativistically, that is, with velocities that begin to approach the speed of light. And when electrons move relativistically, they emit not only the ground tone of the cyclotron radiation, they also emit overtones, and high overtones—high multiples of the frequencies one would expect. In these high overtones there is much more black-body radiation, and too much radiation is emitted; infrared radiation, and

---

[2]The term *black body spectrum* must puzzle the uninitiated. A black body absorbs all the radiation impinging on it. But according to equilibrium theory a black body (but not a black hole, which is not in equilibrium) must eventually emit what it absorbs. In the end the black body radiates what would be produced by a proper final distribution of energy between motion of particles and the various forms of light.

partly even red radiation can be emitted. For a workable fusion reaction to be achieved this radiation must be reflected back. This procedure is somewhat difficult but not impossible. Incidentally, when the Russians published their results in *The Physical Review,* the first sentence began, "The revolutionary movement of the electrons . . . ."

Russia had indeed been moving very fast and very effectively. Meanwhile, we encountered many difficulties, mostly in the form of instabilities. Our magnetic bottle became leaky. I can give you an illustration of such an instability. At the Livermore Laboratory we pursued one particular way to confine the plasma: we used a straight cylinder with both ends compressed; the magnetic lines were closer together at the ends. Spiraling electrons and nuclei then are reflected at the end, except for those that hardly spiral around the lines but move almost parallel to them. These latter particles get lost; the rest are reflected and remain in the bottle. Incidentally, we now believe that this arrangement could result in smaller fusion reactors, which could be established sooner and for less money. The plan is called the *"Magnetic Mirror"* approach. In a way it results in a magnetic field similar to that of the earth, which is strong near the poles but curves out into space between them. Electrons trapped in the earth's field approach the earth near the poles, producing the display of the aurora borealis.

This magnetic bottle is unstable because the magnetic lines are bunched near the ends and bulge in the middle. The magnetic field contains the plasma as long as the plasma pressure is quite low, but only in case of much higher plasma pressures can we expect substantial ther-

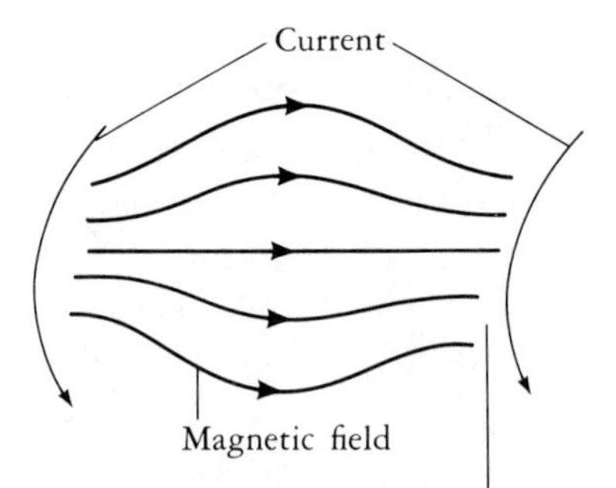

FIGURE 11-3 *The magnetic mirror, an open-ended arrangement. The magnetic field is shown as a bulging bundle from left to right. The curved arrows at the ends represent currents in external conductors that give rise to the magnetic field.*

monuclear reactions. The magnetic lines behave like rubber bands. They try to snap inward and let the plasma leak out between them. This would not happen if the arrangement were completely smooth, but nothing is completely smooth. If a little uneveness exists it will grow in an exponential fashion. The Russians solved that problem. Actually, the theoretical concept underlying the solution arose both in the United States and in Russia, and perhaps in other places. The Russians were the first to demonstrate it.

Their successful arrangement is called the *Minimum B field.* Simply compressing the ends of a cylinder would be foolish. This causes the dangerous bulge in the middle. The compression can be accomplished in a much gentler fashion. One end of the tube is compressed into a thin ellipse, and the other end into a similar ellipse with its long axis perpendicular to that of the first.

Experiment with a sheet of paper rolled into a simple circular cylinder. If you compress the ends into smaller circles the paper will crumple (which is of no particular importance) and the central portion will bulge. But if you shape each end into a flat ellipse, holding the ends between your fingers, with one ellipse compressed horizontally and the other vertically, you will see that there will be no bulge and the paper now does not crumple; the magnetic lines can be imagined to follow the walls of the paper tube and tend to remain straight. (See Figure 11-4.)

We can even make arrangements where the magnetic lines will curve inward rather than bulge outward. The cross-sections at the two ends are, nevertheless, small compared to the cross-section in the middle. The middle will have the weakest magnetic field because at that point the magnetic lines are farthest apart. The mirror works as long as the magnetic fields are strongest at the ends. Yet the instability is avoided.

Today there are two basic approaches to the magnetic bottle proposal. One is the open approach just described. The other is the closed approach. At Livermore we are working with the open approach. In order to further decrease loss through the ends of the magnetic bottle (in our special case through the two ellipses shown in Figure 11-4), we plan to elongate the arrangement into a long tube and put stoppers at the two ends. The stoppers themselves are magnetic mirrors. They are not tight, but they can significantly decrease the amount of escaping plasma.

The closed approach has a modern form in which the magnetic lines of force neither follow the big circle nor the small circle in the torus, but

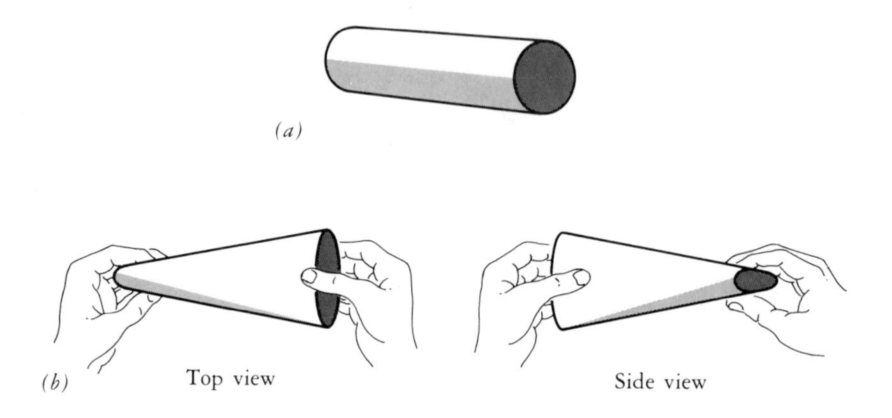

FIGURE 11-4 *(a) A paper cylinder representing a tube with magnetic forces running through it. (b) The cylinder with its ends flattened into ellipses that lie perpendicular to each other.*

instead spiral within the torus, reminiscent of a "braided" French doughnut. If this spiraling is quite gentle, not so steep as suggested by a French doughnut, obvious instabilities can be avoided. (This is done by causing a relatively weak current to flow along the big circle of the torus. The main portion of the magnetic field is caused by strong currents flowing through wires wound around the outside of the torus.) This is the famous Russian design called the Tokamak, the one most people bet on today. From an inferior position in 1958, Russia, through ample funding and hard work, moved into a position of leadership. Tokamak is also pursued with vigor in the United States, particularly at Princeton, where the first doughnut-shaped plasma containment was worked out.

I have mentioned here only two examples of instabilities: the instability in the magnetic pinch due to nonuniform contraction of magnetic loops, and the instability in the Magnetic Mirror. There are many more. But here is a remarkable point. For many years theorists and experimentalists found more and more instabilities, but for more than ten years now we have found no new ones. Perhaps they have all been found. We also learned how to avoid them, or how to decrease their effects to a point where we can live with them. There is reason to hope, but only to hope, that designs now in existence will result in a controlled thermonuclear reaction in the magnetic bottle. I hope this may happen in 1980 or not much later.

One reason for this hope is that excellent practical methods have been developed in Berkeley to inject plasma into the magnetic bottle. Of course, the plasma will not stay indefinitely in any bottle, so an easy way to inject and to reinject is important.

The difficulty with injection is that the magnetic field not only prevents the ions and electrons of the plasma from leaving the bottle, but the magnetic field equally prevents ions and electrons from entering. What has been done, therefore, is to inject neutral atoms that combine the positive ion and the electron; they are not deflected by the magnetic field. The neutral atoms then should be torn apart into nuclei and electrons within the magnetic bottle. This happens in collisions between the neutral atoms coming from the outside and electrons or ions already present in the magnetic bottle. To fill the magnetic bottles with atoms or with the help of electric discharges and currents yielding a low-temperature dilute plasma (ionized atoms), is not difficult. The difficulty is to increase the density of high-energy ions.

The way to turn the above description into reality is somewhat devious. Atoms are not easily accelerated. Therefore, we first ionize the appropriate hydrogen isotopes, deuterium or tritium. The ions can be accelerated with the help of electric fields to an energy of approximately twenty thousand volts, a little less than one two-hundredth of light velocity. When such ions pass through a thin foil they pick up electrons from the atoms in the foil, even if the foil is so thin that the ions are not slowed down to any great extent. Now we have heavy hydrogen atoms of high velocity. These enter the bottle, getting ionized and trapped by the magnetic field. The trickiest part of this procedure is to effect the changes from the neutral atomic state to the charged ionic state and back. We should note that at the low plasma densities in the bottle, the high-temperature ions will practically never again unite into atoms. We should also note that once the controlled fusion reaction proceeds in the bottle, lost plasma can be continuously replaced by the injection of neutral atoms.

None of this research means that we now have something that can be engineered, something that will solve the energy crisis. When the pure research ends in a clear-cut demonstration of an "electric profit," more current output than current input, then the real engineering difficulties start. There are many of them. Why do we not solve these engineering problems now? To some extent we try. In the 1950s, at the very beginning of controlled fusion research, I chaired a steering committee on which Bill Brobeck, an excellent engineer, was serving. Most of the time he was silent. But once, when we discussed the actual engineering problems, he spoke up. "You sound to me like the cavemen who were sitting around a fire they had just learned how to control. One of them said, 'This fire is just fine, but how are we going to use it in a locomotive?'" I think we are a little closer than that to the engineering of controlled fusion, but Bill had a point. First we should find out *what,* and then worry about *how.*

Here, at any rate, are some of the engineering difficulties with the magnetic bottle approach:

☐ If atoms heavier than hydrogen get in from the walls they will emit enough radiation to cool the plasma and stop the reaction. Methods exist for getting rid of these impurities, but we do not know how effectively they will work.

☐ Every thermonuclear reaction emits a lot of fourteen-million-volt neutrons. In the long run they tend to destroy almost any material. We do not know what their action will be on the walls.

☐ To establish magnetic fields in the bottle we need currents that dissipate energy. To get rid of this dissipation it has been proposed that we use superconductors, a peculiar state of metals in which, remarkably enough, no dissipation occurs; currents once established keep flowing forever. Unfortunately, the superconducting state requires an extraordinarily low temperature, near absolute zero. Radiations, particularly fourteen-million-volt neutrons, will get out of the thermonuclear reaction zone. They will impinge on the superconductor, heat it up, and turn it into a common conductor that dissipates energy. To prevent that, the superconducting coils would have to be continuously cooled. This constitutes a further problem at very low temperatures.

☐ The easiest thermonuclear reactions will occur between deuterium and tritium, but tritium does not occur in nature. It has to be reproduced in a nuclear reaction where neutrons impinge on lithium to produce tritium. Most of the energy of the thermonuclear reaction appears in the end as heat in the lithium. The problem is that, at least in the first practical designs of the thermonuclear electric generators, heat engines would have to be used to convert the heat energy of lithium into an electric current.

☐ Some designs of thermonuclear electric generators seem to become big if electricity generation is to be economical. They could become ten, even thirty times bigger in electric output than current fission reactors. The magnetic mirror machines in Livermore may be an exception—provided they succeed. It is hard to utilize too much electricity.

These engineering problems appear much more difficult to me than those we had to overcome in developing fission reactors. And it took twenty years before a reactor could be made for a reasonable price without twenty Ph.D's sitting on top of it. We may hope that we have meanwhile become more clever. So I still say it will take twenty years, and I do not think we can count on controlled fusion before the end of the century.

While pure controlled fusion still appears to be in the future, even in the distant future, there is a possibility that looks truly hopeful within a shorter time. It is the possibility of marrying fusion and fission.

In a truly good match the partners should be different. Fission is strong on energy, but it may become short on fissionable materials. Fusion is relatively weak on energy, but when it proceeds it can produce lots of neutrons that can be used to provide fissionable substances. Therefore, the two approaches complement each other.

Specifically, one may think of a long straight tube in which the plasma is contained by straight parallel lines of magnetic forces. At the end of this tube we could have a system of magnetic mirrors. Fusion could be maintained easily in this system if we demand little or no gain in energy, but want only to have lots of neutrons. These could impinge on common uranium, or thorium, or both. In this way energy will be generated in a manner that can be very safe. What is even more important, we could establish a fuel factory to produce fissionable plutonium or uranium 233 to be burned up in more conventional reactors.

Having discussed the magnetic bottle, let us now turn to the alternative suggestion. Instead of working with a very dilute plasma, let us work with very dense plasmas. This approach is best known as *laser fusion.* Lasers are very intensive light beams by means of which we can concentrate great amounts of energy, thousands of joules[3] or even a million joules in the space of a cubic centimeter or less. Instead of light we may also use electron beams or beams of heavy ions to concentrate the energy.

As yet we have not concentrated enough laser energy into an appropriately small volume. It will be difficult to do. But if we can do it we will produce electric fields much stronger than that holding electrons in orbit within the atom. Such fields tear electrons out of their orbits and will produce instant plasma and instant evaporation of the surface. The force of

[3] A joule is the absolute meter-kilogram-second unit of energy equal to 10^7 ergs, or approximately 0.7375 foot-pounds (if anyone is still interested in these antediluvian units).

the reaction from this evaporation results in the implosion of what is left, so you can hope to compress thermonuclear fuels to a high density.

With the help of these very dense thermonuclear fuels we can produce a miniature hydrogen bomb. The conventional name for such a process is a *microexplosion.* Let me try to explain, without any formulae, why dense fuel can give a small explosion.

The explanation is based on a similarity theorem, which is not precise, but almost precise, and it is exceedingly useful. Assume that we have compressed the fuel a thousandfold. Let us compare the fuel in its compressed state with fluid of normal density a thousand times greater in its linear dimensions. A similarity relation is a relation between two situations; the one is a thousand times denser and one-thousandth the size of the corresponding "normal" one. In the two situations that we are comparing, all temperatures should be the same and all hydrodynamic velocities should be the same. The sound velocity, of course, will be the same.

Consider one deuterium nucleus. Since the density is a thousand times greater, the length of time before this nucleus reacts is reduced by a factor of one thousand. But since linear dimensions are also reduced by a factor of one thousand, the amount of time for the whole assembly to move apart is also reduced by a factor of one thousand. You can see that all the reactions that depend on a collision between a pair of particles will proceed in a similar way because in the comparison between reaction time and expansion time, both have been equally reduced. The course of the reaction is the same, with the "insignificant" change of a thousandfold contraction of time. The theorem, therefore, seems to be correct. If you have one-thousandth the radius you have one-billionth the volume, but at a thousand times the density. Therefore, we have 10^{-6}, or one-millionth of the mass, and the energy released will also be one-millionth. In both cases the reaction will go more or less to completion. So instead of producing energy of a megaton, you produce only one ton TNT equivalent.

Even that is too much. You would like to compress more, to have even smaller explosions. To get an engine going we have to perform microexplosion after microexplosion, probably several times per second. This would be the thermonuclear equivalent of an internal combustion engine. This is the idea of laser fusion.

As I said, the theorem is not quite precise. It is almost right, and to the extent that it is wrong it doesn't matter; but as a theoretical physicist I feel the responsibility to tell you why it is wrong. It is wrong for two reasons.

One imperfection is connected with the slowing down of the fast charged particles produced in the explosion. These particles slow down because of their collisions with electrons. Many such collisions are needed. So long as we can consider these collisions one at a time everything is fine, since each is a collision of two particles. But if you look into the details of the slowing-down theory, you find that the formula for the slowing down has a logarithm in it resulting from the fact that particles at great distance do not interact. There is a shielding effect. This shielding depends on the interaction of several particles, and here the similarity relation does not hold. It holds only if you assume that the logarithm is a constant. Well, a logarithm is almost a constant, so we are not too badly off.

The other reason why the similarity is imperfect is that there is an important three-body reaction, the collision of an electron, a nucleus, and a light quantum. The collision can lead to the absorption of that light quantum and thus prevent energy from escaping. This reaction improves matters, and the greater the density the more the improvement. Therefore, when we make estimates on the basis of the similarity theorem, we really have been too pessimistic. Light quanta may be absorbed in the three-body collisions. Of course this is more important in dense plasma, or fuel, than it would be in dilute plasma.

Returning from this little excursion into theoretical physics, I want to tell you where laser fusion now stands. Everything is fine except that we do not have the lasers, and we do not have the compression. At this writing, we have compressed thermonuclear fuel to little more than ten times liquid density, and produced in one implosion a little more than a billion (10^9) reactions. We shall succeed if we compress to more than a thousand times liquid density, and if we produce in one act at least 10^{18} (a billion times a billion) reactions. The lasers will come. They are being improved rapidly. But it will be quite a trick to develop the exceedingly powerful lasers that have to deliver their energy in a time comparable to the period in which light moves through a centimeter, delivering to a very small volume considerably more than a thousand joules of energy. That is one difficulty.[4] A step toward this difficulty was the construction, under the direction of John Emmett, of an enormously complex apparatus (shown in Figure 11-5) named after Shiva, the Hindu god of love and destruction.

---

[4]Lawrence Livermore Laboratory achieved an initial success in the middle of November 1977. It took a gigantic effort in money and diligence.

The other difficulty is to produce a thoroughly symmetric implosion. I believe that in time both of these difficulties will be overcome.

I have told you that after lots of arguments we reached the point where we could talk with complete freedom about magnetic bottle fusion. We should be allowed to talk freely about laser fusion, but we aren't. Official secrecy still engulfs us. Therefore, I cannot tell you all the reasons for my optimism. I can tell you that my optimism exists. My optimism extends to the belief that in the next few years we will obtain "energy breakeven" in laser fusion, meaning that we will get out of the thermonuclear reaction as much energy as we put in as laser energy.

That amount, of course, is not enough, because laser energy is very expensive. Lots more energy yield is needed to make laser fusion pay.

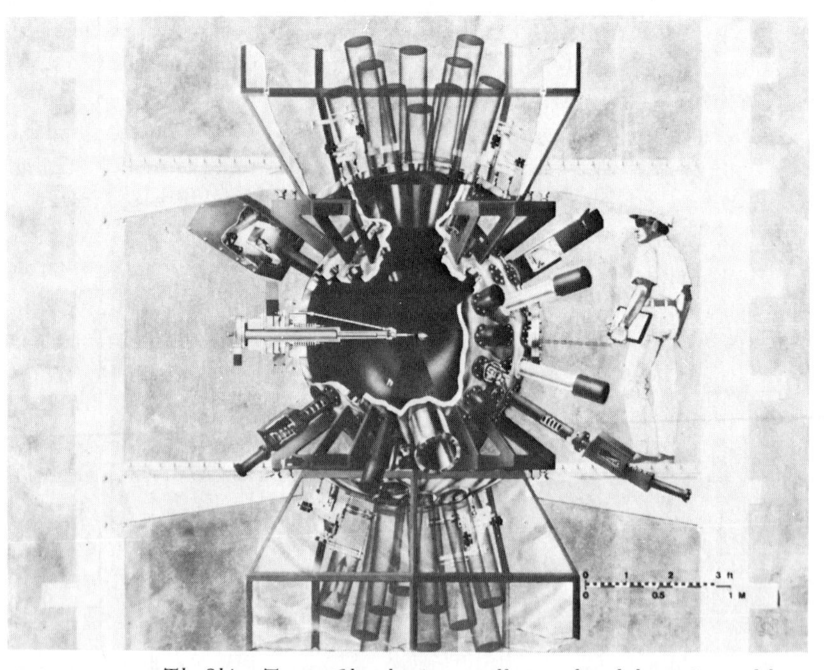

FIGURE 11-5 *The Shiva Target Chamber is a small part of needed experimental laser fusion apparatus. Laser light is brought in simultaneously from ten pipes on the top and ten pipes on the bottom. Compression and nuclear reaction occur in a tiny dot at the middle of the sphere. Apparatus practically filling a whole building feeds the twenty pipes, or the arms of the god Shiva. According to Hindu creed, Shiva had three eyes: two for seeing, and one (usually kept closed) to emit annihilating radiation. The Hindus obviously knew about lasers.*

Furthermore, even if we could maintain a reactor, think of what would be involved. We would need an apparatus that could work for years and survive billions of microexplosions, which are not so "micro" as the explosions in the internal combustion engine as we know it today.[5] Slightly radioactive surroundings would be produced, so if something went wrong we could not get in there with our fingers and fix it. The entire mechanism would have to work reliably and inexpensively. The engineering job will take a long time, and we cannot be sure that it will succeed in a proper economic sense.

Some of my close friends who are great enthusiasts are telling me, "Laser fusion is the way. Look how far we have come in just a few years." Of course, they have accomplished a lot. But I would remind you of Niels Bohr's definition of an expert. According to him an expert is someone who, by his own personal, painful experience, has found out all the mistakes that one can commit in a very narrow field. In controlled fusion I believe that no experts exist. The people working on the magnetic bottle have almost become experts, though. They have made lots of mistakes, and have learned from them.

The people working on laser fusion have the pleasures of making all those delicious mistakes still ahead of them. When they succeed—and I prefer to say "when," not "if"—I believe that a lot will have been learned. Ways will have been found to compress material that my grandfather and even my father considered incompressible. We shall compress these materials a thousandfold and do it out in the open, so that for a fraction of a nanosecond (10^{-9} seconds) we will be able to look at it. That will be very interesting physics.

Anther reason why laser fusion or any controlled fusion may not solve the energy problem is not that it is impossible, but that it may just be too expensive. The dollar break-even will come much later than the energy break-even.

But there is one region in which the dollar has a different kind of value. If you want to go into space travel, you may be willing to settle for

---

[5]The chambers containing these microexplosions will probably have to be massive metal chambers several meters in diameter. We would have to protect the walls from all effects of the explosions; maintain the lasers and the path of the laser light into the cavity; and, at least once a second, drop very small pellets containing the appropriate thermonuclear fuel into the focus of the laser, with high precision and through openings that would have to be closed during the explosion itself. Such requirements will be a nightmare for the uninitiated engineer who first encounters them. How it will look in the light of engineering experience no one knows.

an apparatus that works for only two months. And you may be willing to pay for it many thousands of times the amount you would pay for a terrestrial power plant, because you would be getting a machine that could emit a very fast jet, and therefore could accelerate without using a great mass of fuel. We may get to the moon, or for that matter to any place in the solar system, without a rocket taller than the tower of Babel.

I believe that all these possibilities in science and technology are right ahead of us. Controlled fusion, therefore, is not only a difficult problem, and as interesting as all difficult problems are. It is also fascinating because it can lead to so many things. In energy production it opens up the use of fuels that exist in great abundance. Of deuterium we have as much as we like. Tritium we can make from lithium—practically unlimited quantities may or may not be available, but there certainly is a lot of it. Furthermore, it would give us a nuclear machine that generates fewer kinds of radioactivities and, even more important, generates less dangerous radioactivities. Controlled fusion has the agreeable property that when it malfunctions, it stops. These are some reasons for the great popularity of controlled fusion.[6]

There is one final reason for that popularity: ignorance. Many people who advocate controlled fusion do not understand how difficult it is. They are not the experts, and they promise more than can be delivered. They deflect attention from the really feasible ways in which the energy crisis can and must be solved. This ignorance is further fed by secrecy. If it is difficult to explain to a senator why an apparatus won't work because of intrinsic complications, it is more difficult still to explain to him why it won't work because it's a secret. This problem can be overcome by obtaining security clearance for the senator. The poor senator has no time to

---

[6]There is one proposal which is highly *un*popular though not necessarily unrealistic. Small fusion explosions can be produced today in a relatively inexpensive manner. They run on deuterium, which is practically inexhaustible since it can be obtained from the ocean. The explosions could be carried out in salt cavities while the energy is absorbed in appropriate water or carbon dioxide vapors within the cavities. These reactions could produce plentiful energy, and, what may be much more important, they could transmute thorium into useful fission fuel.

The project is difficult mainly because it requires that the cavity should stand up under thousands of explosions performed probably more than once a day. Its great unpopularity is due, of course, to the fact that anything directly connected with nuclear explosions inspires fear. Obviously, if such a project is ever to be implemented, its safety must be subjected to the most careful scrutiny.

listen. You have to clear his administrative assistant. You have to clear the constituents who write him letters. So long as we have secrecy we cannot have general understanding. Secrecy and science are not compatible. Secrecy impedes progress. Ignorance is bad enough. Ignorance due to secrecy may be incurable and eventually fatal.

CHAPTER 12

NATURAL ENERGY SOURCES

In which uncertain predictions are made about an uncertain future. Solar energy will come, but how? The heat of the earth will be used, but where? There are other possibilities, but which?

All energy sources are, of course, natural. Today, however, sun and wind and waves and hot springs are somehow considered more natural than coal or oil or nuclear energy; objections to the latter three have been raised for various reasons. In fact, all of the living world depends on solar energy. In many cases animals and birds follow the sun. Man's first application of solar energy was in the form of firewood. The sun satisfied the most basic needs of agriculture and, for this reason, in some early communities it was watched and studied with extreme care. Preindustrial civilizations used these natural sources in sailing ships, windmills, and waterwheels. The rays of the sun were used to lend occasional brilliance to the beautiful windows in cathedrals.

The newly popular energy sources discussed here are often called renewable. Some of them are. Readers will find, perhaps to their surprise, that some of them are not. To harness them effectively and adapt them to modern conditions present a problem. It used to be that America could solve all problems by employing a magic wand: technology. In the view of many young people that magic wand has turned into a witch's broom, and on this frightening device we are collectively riding to disaster. I tend to believe the older myth, but with limitations. The magic wand works but it does not work instantaneously.

In the case of energy there are a great number of real possibilities awaiting development. An even greater number of energy proposals are completely unrealistic. But we can feel reasonably sure that energy will be available, even easily available, in the long run. It is most important to remember, however, that miracles of technology and industry are not instantaneous. No matter what we do the next years will be hard, not only for selected countries but for most of the world. One of the most important points in every single case is to ask when an energy source can become available. Timing seems to be at least as important as the anticipated quantity.

Solar energy is truly the oldest form of energy. There is a good chance that it will turn out to be the most important answer to our energy problem. It is ample, and as human lives are measured it is everlasting. The sun has been shining for five billion years and there is no reason why it should not continue for five billion more. In this sense it comes closer to being an inexhaustible source than any other practical possibility.

Because the bureaucratic machine in Washington does so for purposes of federal funding, it has become customary to include in the general category of solar energy all energy sources that continue to be replenished by the sun at the rate at which they are actually used. Therefore, when we speak of solar energy we mean not only the collection and utilization of sunshine. We also mean exploitation of warm surface water in the ocean, called Ocean Thermal Energy Conversion, which is difficult; we mean the more hopeful use of vegetation that derives its energy from the sun. We further include windmills driven by the atmosphere whose movement is caused by the sun. We include the exploitation of the ocean waves that have been whipped up by the winds. We include hydroelectric power, which depends on rainfall caused by water evaporated by the sun. We do not include fossil fuels that have been derived from the sun because they are being used up much faster than they are being replaced.

Solar energy has become exceedingly popular because it is conceived as clean and reusable, and because work on it has been neglected in the past. That this deficiency is being remedied is obvious from Table 12-1. Of course, expenditure of money on research is necessary, but not sufficient of itself. It remains to be seen what the actual prospects are for each of the various forms of solar energy.

We start with the direct collection of solar energy. As stated earlier, the greater the temperature difference, the more heat energy is usable. Conversely, gentle heating is more easily accomplished than heating to extreme temperatures, particularly when energy is delivered directly by the sun. Applications range all the way from heating a swimming pool where a temperature increase of five to ten degrees Centigrade affects its usefulness, to concentrating solar energy into sharp focus to produce ex-

TABLE 12-1 *U.S. Appropriations for Solar Energy*

Fiscal year	*Amount appropriated*
1971	$ 1,230,000
1972	1,680,000
1973	5,080,000
1974	17,280,000
1975	41,900,000
1976	114,600,000
1977	290,400,000
1978 proposed	305,000,000 plus

tremes of temperature. In fact, temperatures comparable to those prevailing on the surface of the sun can be approached. These are generally destructive, but they can play a role in special industrial applications.

What is easiest will come first. We should therefore begin by considering cheap methods to accomplish gentle heating. The general approach is to utilize what is called the *greenhouse effect.* This is done by admitting sunlight through glass. Then it is absorbed inside the enclosure, by a black surface, for instance, and we take advantage of the fact that glass does not easily allow heat radiation to escape.[1]

For those to whom this procedure is not obvious and familiar, it should be said that the temperature of our planet is regulated by an inflow and outflow of radiation. The inflow, of course, comes from the sun. The outflow is due to radiation similar to light, except that this *heat radiation* has a longer wavelength. A closer look at such equilibrium explains some phenomena with which most of us are familiar.

It can get very cold at night in the desert. There is little moisture in the air and moisture impedes the outflow of heat radiation. The desert, lacking a blanket of moist air, cools rapidly. A similar situation is discussed in Chapter 7, where we consider carbon dioxide as a heat blanket.

It is widely believed that the cold weather brings snow in winter. In some respects it is more correct to say that snow cover brings about a succession of cold days or weeks. Indeed, snow reflects sunlight, so little of the inflowing radiation is retained. At the same time, snow is very good at emitting heat radiation. Yes, even snow has heat, and what heat it has is easily emitted. Therefore, once a big area is covered by snow it tends to remain cold.

What happens to climate on a big scale is reproduced on a small scale in the solar greenhouse. Since glass hinders the escape of heat radiation, it will be hotter in the greenhouse, and this heat can be utilized not only for growing plants but also as an energy source.

---

[1]Israel has practiced an ingenious variant: the solar pond. Instead of glass, water itself is used, which of course is cheaper. Sunlight is absorbed at the black bottom of the pond, and the water impedes the escape of heat radiation. The bottom of the pond, however, will become hot, and convection in the water will lead to increased heat losses. Therefore the solar pond is layered. The bottom is highly saline, and consequently dense. Intermediate layers have less salinity, and the top consists of fresh water. By this rather cheap method one can obtain almost 100°C at the bottom. Unfortunately, the overall efficiency of the arrangement turns out to be not much more than 1 percent.

The standard application is the production of hot water, which has been practiced for many years. This use is particularly popular in Israel and the island of Cyprus. It is more effective when there are many cloudless days. In the United States such heat collectors are not limited to the South; they also can be used to advantage in the Rocky Mountains, where there is a lot of sun and a lot of cold weather. In the Rockies, however, water alone cannot be used in solar collectors. On too many occasions it would freeze. The collector should contain water and an antifreeze in solution, or air that is brought into contact with water after it is heated. Hot water is useful by itself, and can also be used to heat a house during an entire cold period, including the cold nights that may continue into the summer.

Availability of solar energy to various parts of the United States is shown in Figure 12-1. The South, of course, receives more sun than other areas. The cloudless desert of Arizona receives the most, and with the greatest number of sunny days coupled with open space, Arizona has the greatest potential for solar energy utilization. Clear skies over the Rocky Mountains help to extend available solar energy toward the North despite colder temperatures.

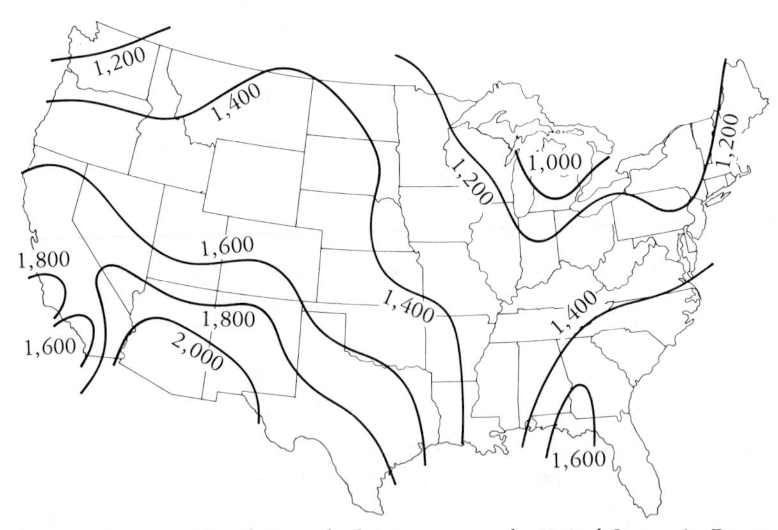

FIGURE 12-1 *Distribution of solar energy over the United States, in Btu per square foot per day. (Justin A. Bereny, comp., and Francis deWinter, ed.,* Survey of the Emerging Solar Energy Industry, 1977 Edition *[San Mateo, Calif.: Solar Energy Information Services, 1977], p. 10.)*

It should be pointed out that in an area with 1,200 British thermal units per day solar energy would be far less practical than in those areas with more sunshine. The figures are averages, and in those more northerly regions there are not enough sunny days during cold weather.

Greenhouse heat collectors, incidentally, work even on cloudy days. The sun's rays are less direct then and they are decreased to some extent, because clouds scatter much of the sunlight away from the earth even before the light reaches its surface. Nevertheless, the heater still works when skies are overcast, though it does so with less efficiency.

Flat plate collectors were popular in Florida in 1950, for producing hot water. Competition from cheap natural gas displaced these collectors, with the results shown in Figure 12-2. Now, with rapidly rising gas prices, the solar collectors should be installed for good.

The greenhouse principle was turned into practice years ago. But we must ask whether it is economical. The main difficulty is the cost of installation. As a general rule, mass production lowers costs, not only for automobiles but also for solar collectors. If big quantities are at stake, production methods can be optimized and costs due to variations and modifications avoided. This principle is known as the *economy of scale.* But solar energy utilized at each individual dwelling, which is only partly compatible with the economy of scale, may still be the right choice for a particular reason. Massive production of energy is cheap at the big power stations where electricity is generated, but considerable amounts of money

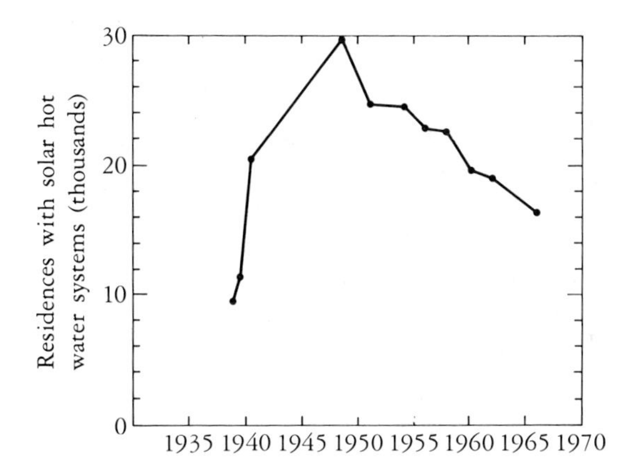

FIGURE 12-2 *Solar water heating systems in operation in southern Florida. (Bereny and deWinter, p. 47.)*

must be spent to distribute the energy. Solar energy is delivered free of charge to individual small homes. For a forty-story building, solar energy is less advantageous. The big volume of a skyscraper has relatively little surface on which to collect solar energy.

Simple flat plate collectors of solar energy still cost too much for widespread use in the United States. Though common, glass is expensive. There is hope of replacing it by cheaper synthetics that function similarly to impede the outflow of radiation. Unfortunately, present synthetics lose their effectiveness under the influence of the ultraviolet part of solar radiation. A relatively inconspicuous development, such as the discovery of cheaply manufactured sheets that will not deteriorate and that can be used as greenhouse windows, could make a great contribution toward better utilization of solar energy.

It has been mentioned that one-sixth of the fuel in the United States is spent on producing process steam for industry. One point was that considerable energy is used up in boiling water, and with little additional fuel much higher temperatures can be achieved.

It is advantageous to generate process steam with most of the energy coming from the sun. In a well-constructed collector of the greenhouse type, water can be brought to, or almost to, the boiling point. If the boiling point is not quite reached, a small reduction in pressure will cause the water to boil. This evaporation process requires the most energy, and it can be accomplished by the relatively low temperatures near the boiling point of water, which can be obtained from the sun without complex apparatus.

The steam, however, is now not hot enough. Its temperature can be raised as necessary by compressing it, without any heat transfer. The work of compression turns into heat. At first this seems to be the wrong way in which to proceed, because mechanical energy or compression energy is the most expensive form of energy. But if most of the energy comes from boiling the water, and only an added 10 percent is needed to compress and heat the steam, we can afford to use simple electrically driven mechanical apparatus to perform the compression. Good electrical generators work at 40 percent efficiency and therefore need 2.5 times the fuel energy delivered as mechanical work. But if this mechanical work makes up only 10 percent of the process steam energy, we have used only 2.5 times 10 percent, or 25 percent as much fuel as we would have used had we simply boiled water to get high-temperature steam with 100 percent efficiency (which can never quite be reached). The rest of the energy is obtained from the sun.

Solar energy could be used to much better advantage if higher temperatures could be reached. One step in this direction, using the greenhouse effect, would be to drastically reduce the heat lost through conduction. A proposal for accomplishing this end is to collect the sunlight in a double-walled glass vessel, with a vacuum between the walls as in a thermos bottle. In that case the only way to lose energy is by the heat radiation which we have already discussed, and we would have almost entirely eliminated heat transfer by materials. It is possible, but by no means obvious, that these more ideal greenhouse collectors could be produced cheaply enough to make an impact.

The direct consequence of the higher temperatures would be utilization of solar energy not only for heating but also for air conditioning and refrigeration. That heating in a solar collector should result in the cooling of a house sounds like a miracle. It is. Yet it may be barely possible to explain it so that the readers imagine they understand it.

This miracle is based on principles of physics, more specifically on the principles of thermodynamics that we encountered in Chapter 5. What counts are temperature *differences.* If we create a temperature difference between the hot air outside a house and a hotter place, the solar collector, we can "transform" it into another temperature difference between the hot outside and the intended cooler temperature inside the house.

There are many ways in which this miracle can be performed. What I shall describe is practical, and a good approach, simplified in order to avoid cluttering the principle with details. The method requires a fairly common malodorous gas, ammonia, and a greater amount of water. It is done in an airtight system so the ammonia cannot offend our sense of smell.

Ammonia readily dissolves in water, and in so doing it cools the water. That may seem somewhat strange. If heat is not freed when the ammonia gas dissolves then it appears that the water does not attract it, so how is it that it dissolves so readily?

The answer is not quite obvious. Most of the ammonia that enters the solution, together with the water, produces two charged molecules called ions (mentioned later in this chapter when we discuss fuel cells). The ions carry opposite charges and attract each other; energy is needed to tear them apart. In spite of their attraction they move apart or, to use the technical term, *dissociate,* because—and this is not just a figure of speech—they like the freedom of independent motion.

These are the results: when ammonia is dissolved the solution becomes cool; when ammonia leaves the solution the material is heated. Conversely, at higher temperatures ammonia tends to leave the solution; at lower temperatures more ammonia enters the solution. But we should note an important circumstance: the hot ammonia above the hot solution, while less in density, is greater in pressure than the cool ammonia over the cool solution. This is not so remarkable; hot gases at equal density have a higher pressure than cool gases.

Now what happens is pretty straightforward. The solution heated in the solar cell emits ammonia. The water is hot and reduced in ammonia content; the ammonia gas is hot and at high pressure. The two can be driven separately to a new cell where they will again come into contact. But before renewed contact is established, both are lowered in temperature from hot to merely warm, being cooled by the air outside the house. Then the ammonia is dissolved by the water, which brings the temperature inside the house below that of the outside air, and air conditioning is thus accomplished. From there the fluid proceeds to the solar cell and the cycle can start anew.

The arrangement can be worked out so cleverly that no pump is needed. The solution and the gas flow into the right places without assistance. This is the magic of thermodynamics, physical chemistry, and engineering. It works, provided we can get the blessing of economists. This is not easy because high-temperature solar cells are needed and they are expensive.

The usual approach toward solar energy of higher temperature, and therefore higher quality, is to collect solar radiation by mirrors. In Figure 12-3 we see a printing press operated for publication of *The Sunshine Journal* in France in 1878, powered by a two-thirds horsepower solar steam engine. Here a parabolic mirror, together with the power plant, has to be moved to track the sun. In modern proposals the parabola is broken up into plane segments located on the ground. Their orientation is changed in such a way as to collect solar energy in an unmoving focus.

In the greenhouse effect we usually are able to raise the temperature by a relatively small amount, mainly by cutting down losses due to heat radiation. These losses increase very rapidly if the temperature becomes higher. Therefore, it is difficult to raise the temperature by a factor two on

FIGURE 12-3 *In 1878 a printing press to publish* The Sunshine Journal *in France was powered by a two-thirds horsepower solar steam engine. (Courtesy of Solar Energy Information Services.)*

the Kelvin scale. For example, going from room temperature, or 300°K, to 600°K, the loss due to heat radiation increases sixteenfold, which is 2^4, or two times two times two times two. It follows that if we can concentrate solar radiation sixteenfold by the use of mirrors, and not do anything about impeding outgoing heat radiation, we get an equilibrium temperature of 600°K. At this point generation of electricity by use of a solar steam engine will look promising.

In reality the situation is much more complicated. In arriving at 300°K, for instance, we have already taken the average over day and night when balancing incoming and outgoing radiation. A solar engine will, of course, work only in the daytime, and at least a factor two is therefore available even without mirrors. With ample use of mirrors one can get to very much higher temperatures.

There is a reverse side to this effect. The full power of the sun is available only part of the day. The situation is particularly poor on cloudy days when concentration by mirrors will not work. Therefore, solar electricity plants are best located in desert regions, such as Arizona in the United States, or most of the Middle East.

Mirrors should be deployed that reflect the light of the sun onto a definite spot on the earth's surface. In order to compensate for the motion of the sun, the mirrors have to be moveable.

The standard approach at present is to put the steam electric plant on a high tower. Then plane mirrors can be located on the earth's surface. They have to be moved by clockwork and moved with considerable precision. These specifications result in a plant that is unconventional, expensive, and not easy to construct, even though all the basic principles are simple and straightforward. The expense of storing energy for the night must, of course, be added.

Many storage methods are available. What one may think of first is storage in electric batteries. Batteries at present are expensive and relatively short-lived. Also, they do not return all the stored energy, but convert as much as one-quarter of it into heat. According to present technology, costs of energy storage alone could be comparable to the capital expenditure for both mirrors and the steam electric generator on the tower.

There are concrete hopes of reducing this cost by better batteries. Particularly, we might produce batteries which, instead of lasting for a few years, may last for a few decades.

Another truly exciting possibility is use of the flywheel. Rotating flywheels can store considerable amounts of energy for a period of hours or days. There need be little loss if good bearings are used, and if the flywheel rotates in a vacuum. Unexpectedly, the best hopes for flywheels are for equipment made not of steel, but of appropriate synthetic materials. In each case the stress the flywheel has to bear is proportional to the energy, and it is independent of the material's density. A lighter synthetic, rotating a little faster, will store the same energy and will have to sustain the same stress per unit volume as steel. Both energy and stress are proportional to the density and the square of the velocity. An ideal flywheel should stand up to strong tensions along the concentric circles of the wheel, but at the same time it is best to have a material that is compressed with relative ease along the radii, that is, in directions perpendicular to the circles. Steel behaves the same way in every direction. However, a synthetic with fiberglass embedded along the circles becomes very difficult to stretch along the circular path, but easy to compress at right angles. There is real hope for excellent flywheels from an appropriate synthetic. Its two outstanding properties would be long life and the recovery of 90 percent of the energy deposited.

Energy storage is, of course, a common feature of solar electric developments. But the mirror arrangements that have been described are not without competition. We may, for instance, focus the sunlight along a line rather than toward a point. In such an arrangement we would use a trough-like parabolic structure as a reflecting surface. Focusing can be achieved along a line near the bottom of the trough without the reflecting surface being moved hour-by-hour or minute-by-minute to follow the course of the sun. We have to adjust the trough two or perhaps four times a year to compensate for the sun being higher in the sky in the summer than in the winter. The disadvantage is obvious. We obtain a smaller flux of solar light and a lower temperature than we could get if we collected the rays of the sun near one focal point.

Finally, we can return to effective solar panels, in particular to the evacuated double-glass cylinders. These could generate electricity with little assistance from mirrors, or conceivably even without mirrors.

The lower the temperature of the working fluid, the more attention we must pay to the best possible use of the limited solar heat obtained. A solution in this case is to use an arrangement that is similar to a bottoming plant in an electric generator. Water at low temperatures, even above the boiling point, produces a low vapor pressure and great volumes of vapor. Thus, very big turbines are required to produce substantial amounts of electricity. Capital investment is heavy. Therefore, it is better to use freon or ammonia as the fluid that is alternately evaporated and condensed, because these substances have higher vapor pressures at lower temperatures. But no matter what we do, thermodynamic limitations cannot be exceeded. The best we can hope for is to convert a fraction of the energy into electricity, somewhat less than the limit given by the temperature difference in the process divided by the higher of the temperatures.

Once a solar plant has been constructed, it of course will not require any other fuel to operate. But it still will require maintenance, and maintenance in this case is apt to be more expensive than in a conventional plant. In addition to keeping all boilers, heat exchangers, turbines, and generators in order, we must provide for the cleanliness and good condition of many acres of surfaces that collect solar energy.

The upshot is that proper predictions of the cost of solar electricity turn out to be much greater than the cost of electricity from other sources —coal, oil, or nuclear—if one applies the present state of technology.

Even optimistic projections forsee a price for solar electricity five times the usual cost unless truly ingenious innovations are introduced. Unfortunately, what we have discussed so far is rather conventional technology, in which slow progress is more probable than brilliant new ideas and breakthroughs.

The situation is quite different if, instead of solar thermal electricity, we consider a method that converts sunlight directly into electricity by using photoelectric cells. Such cells have served in our efforts at exploring space. The only trouble is that the cells used in our spacecraft are too expensive to be practical for general terrestrial use. In space exploration we could afford to spend a thousand times more for electric currents than we are prepared to pay in commercial generating systems.

The working of these photoelectric cells is based on the fact that light arrives in packages called *quanta.* Light quanta can impart their energy to electrons, which thereupon will be shoved from one substance into a neighboring substance. The appropriate substances are known as semiconductors. It is not difficult to make arrangements in which the electrons find their way back to the place they came from along an external electric circuit. What is difficult is to reduce the cost of the semiconductor sandwiches by a factor as great as one thousand. What has been done so far is to reduce the cost by approximately a factor ten. The remaining factor one hundred may be more difficult to achieve.

One should consider the problem of solar cells as the question of the mobility of electrons, the extremely lightweight carriers of electricity. One of the most peculiar facts of physics and technology is that electrons move very differently in different materials. In metals they move almost freely; in insulators they move with the greatest difficulty; in superconductors, as mentioned, they move with complete freedom. Our whole electric network is based on the fact that electrons move a thousand kilometers along a copper wire but do not move one millimeter across the insulating wrapping that surrounds the wire. Semiconductors, intermediate between metals and insulators, are the stuff of which solar cells are made. Under varying conditions the mobility of electrons is sharply changed. Sunlight moves them. Small additions of foreign substances influence their behavior. In the past, regular arrays of silicon atoms in single crystals of silicon have been used, together with sharply defined impurities that regulate the behavior of the electrons. Photocells of the past were made with the care required in the manufacture of a Swiss watch.

Can long ribbons of cheap single crystals be grown by mass production methods? Can other substances, such as sulfides or gallium arsenide, be substituted for silicon? Neither silicon nor its substitutes are expensive; the ticklish item is their fabrication. Are crystals really needed, or can we work with the more disorderly array of atoms found in amorphous substances?[2] Can the work be done at higher temperatures, permitting us to concentrate more sunlight on an expensive photocell?

All these possibilities exist. Success is in no way assured. We are faced with the promises and the uncertainties of high technology under conditions where we have partial understanding of a great number of factors. There is real hope for success; there is only limited expectation for a rapid resolution of all difficulties.

Apart from the problem of high costs, photoelectric cells are more strongly dependent on the weather than are simple hot water collectors. The latter utilize all of the sunlight, very particularly the longwave, or red, components that penetrate more easily through an overcast. Photocells depend more on the shortwave, blue and violet, components, which have difficulty getting down to earth even through thin layers of cloud.

Of course, mirrors may be combined with photoelectric cells. This procedure drives up the cost because of the expense of the mirrors. On the other hand, the cost of the photocells will decrease in proportion to the area that must be illuminated. Unfortunately, the output of a photocell does not increase in proportion to the sunlight received. Focusing too strongly increases the temperatures and decreases the efficiency of the cell. In this respect, gallium arsenide seems to work better.

One great potential advantage of photocells is that they deliver small amounts of electricity as effectively as big amounts. The economy of scale does not apply. In sunny climates at some future date, photocells, possibly combined with focusing mirrors, might provide an isolated house with all its energy requirements, including electricity and air conditioning. This system would provide a paradise for the modern hermit who loves his comfort but adheres to the conviction that only what is small can be beautiful.

A reasonable compromise may be to develop solar electricity for

---

[2] A paper by Stanford R. Ovshinsky and Arum Madan, "A New Amorphous Silicon-based Alloy for Electronic Applications," *Nature 276:* 482–484 (1978), points out that effective and cheap photocells could be produced in the near future by employing the adjustable properties of noncrystalline materials.

small communities, based either on steam-driven generators or on photocells. In this way energy storage would not be quite so difficult, because the needed storage devices of intermediate size may turn out to be more economical than small individual units. Distribution of electricity within a small community remains inexpensive.

In general, remote locations are appropriate for early applications of solar energy. In the middle of Africa, on the edge of the Sahara Desert, water is found underground, and the local people spend more than half of their labor bringing that water to the surface. To lay electric cables over a thousand miles is impractical there. To bring in petroleum is not much better. The best alternative seems to be to construct a solar engine, even though the capital investment is twenty or thirty times as high per unit of energy developed as in an advanced country. If the possibility of development is to be opened up in these regions, expensive solar power is still the cheapest. A project of this kind in a former French colony is sponsored jointly by the government of France and American AID.

A more modest scheme of great potentiality could be started in southeast Asia. In these regions, fishing and even the farming of oceanic food are practiced, and could be used to a much greater extent. If the seafood is to be more widely distributed, or even exported, refrigeration will be needed. We have seen that solar collectors of a not too expensive variety can be used for refrigeration. The availability of such refrigeration units on shipboard or at locations where cheap energy is not available could be of great use.

An additional approach should be mentioned, not only because it has attracted millions in research dollars, but also for the reason that it is highly imaginative and therefore amusing. Solar cells have been used in space to produce relatively small amounts of electricity. Could we produce big amounts of energy on a satellite and then transfer it to the earth? A satellite moving around the equator at a distance of approximately fifty thousand kilometers from the center of the earth completes its circuit in twenty-four hours and will retain its position over one point of the equator, appearing to be suspended. It may be possible to deploy solar cells, with or without mirrors, on such a satellite. One might even employ the more primitive method of producing hot gas or steam and using a turbine-generator system. Of course, a closed cycle would be needed, with complete reuse of the working fluid. Successful development of the space-shuttle would be required to lift the mass into such an orbit, at a cost that is exceedingly high but might conceivably be afforded.[3] Some of the equipment, for instance the mirrors, may weigh very little, as long as they are not subjected to any forces, particularly not to any gravitational forces.[4]

It is of course necessary to transfer the energy down to earth. The plan is to transform electrical energy into microwave energy, which is electromagnetic radiation of wavelengths measured in millimeters—that is, shorter than radar but much longer than common infrared or heat radiation. Microwaves penetrate the atmosphere and most clouds with relative ease; they are attenuated by big droplets that occur in heavy rain. On the surface of the earth, microwaves will have to be transformed back into electricity, and the job is then complete.

Two great advantages to this system would be that it could function with considerable reliability for practically twenty-four hours a day, and the photocells could be reached by all wavelengths from the sun, including the violet and the shorter and more effective ultraviolet radiation. The disadvantages are that the system obviously is highly complicated, that space stages are expensive to establish and maintain, and that there may be trouble with damage from high-intensity microwaves. In this case, as in many others, environmentalists may advocate a particular form of energy as long, and only as long, as it appears not to be feasible.

It would seem reasonable to consider the separate portions of the solar-space energy project one by one, and to translate them into practice one by one before attempting to establish the whole complicated system. For instance, we might develop the microwave transmission system over ten kilometers or fifty kilometers before planning to send radiation over fifty thousand kilometers. There are many offshore islands where trans-

---

[3]In one project a mass of 100,000 tons is mentioned, which could produce 50,000 megawatts of electricity (comparable to the amount produced by fifty nuclear reactors). To lift a ton into "synchronous," or apparently stationary orbit, the cost would be $14 million to $20 million. The cost of lifting into orbit the material for generating 50,000 megawatts is estimated at between $1,400 billion and $2,000 billion, to which considerably more must be added for costs of fabrication and maintenance of the equipment needed in space and on the ground. Such amounts can be compared with $50 billion if fission reactors are used to generate the same amount of electricity.

[4]Of course gravitation is there, but it is cancelled by centrifugal force. One of the nice points of Einstein's theory of gravitation is that this cancellation may be regarded, in theory as well as in practice, as complete absence of forces.

mission from the mainland by microwaves might compete with transmission by cable. In this way practical experience could be obtained on a relatively modest scale.

One particular example may be found in Hawaii. The big island of Hawaii has active volcanoes. Great geothermal deposits were found on the peninsula jutting out from Kilauea toward the southeast into the Pacific, apparently more ample than can be used by that island's small population. Electricity could be generated and transmitted by wires to the top of the extinct volcano Mauna Kea. The top of that huge cone, reaching 4,500 meters into the sky, can be seen from Koko Head near Honolulu, with its population of nearly a million. Energy could be transmitted by microwaves from Mauna Kea to Koko Head. If this undertaking proved acceptable and profitable, it might serve as encouragement to pursue the solar space project.

Another step might be to launch big, relatively light mirrors into orbit five hundred or a thousand kilometers from the surface of the earth. Orientation of these mirrors could be controlled, and they would be used to intensify and extend the daily duration of solar radiation by reflecting the sun's rays. Because of the size of the solar disk, the added radiation would be spread over a considerable area—at least twenty kilometers in diameter. The higher the mirror, the larger the area. But solar radiation could be directed toward a city such as Denver, making installation of flat plate collectors more economical. As they circled the earth, the mirrors would be oriented by remote control to bring added sunlight to appropriate and changing areas. If no added radiation were needed, the reflection from the mirror could be directed back into space. Direct application to a big solar electric source does not appear practical.

On the other hand, application of solar-space technology to weather modification, a development which is in the forseeable future, could become most important. The energy-starved "polar front" that separates the winds of the temperate zone from the polar air mass, for instance, could conceivably be influenced. Processes along the polar front seem to determine weather patterns, and can make the difference between a hard or mild winter, between drought or ample rainfall.

All of these possibilities mean that the components of the space-solar system may be more important than the system as a whole. For the near future the solar-electric-space proposal *in toto* appears truly discouraging. It has been discussed at some length here in order to make that point clear. Furthermore, some attempts to develop parts of the system may bear quite remarkable fruit, particularly if we think in decades rather than in years.

Instead of converting solar energy into electricity we can convert it into fuel. Burning of wood is the most ancient form of energy production, and burning of coal or oil, which are fossilized forms of vegetation, has been the most important form of energy throughout the Industrial Revolution. These fuels have been discussed in earlier chapters.

It is not advocated that we return to the burning of wood on a large scale, though it is seriously suggested that we grow wood and transform it into methanol, also called wood alcohol, which is the simplest of all the alcohols, and then use methanol instead of or in addition to gasoline. There is a difference of opinion as to whether or not the cost of methanol derived from wood can be reduced sufficiently to make such an enterprise worthwhile, but the project is bigger and broader than the use of wood. It utilizes what is currently called *biomass,* plants that convert solar energy into fuel via a biological process.

Perhaps the most hopeful direct approach is to grow algae in water and convert them into a fuel like methanol.[5] One advantage of this process is that it could use salt water, which is much more available than fresh water. Another advantage of growing algae may be that the ratio of conversion of solar energy into the energy of fuel could become quite high. Under normal conditions this conversion would be 1 or 2 percent. But if algae are grown in an atmosphere enriched in carbon dioxide, the utilization of sunlight can reach 10 percent.

In Chapter 7 we mentioned that massive burning of fossil fuel will increase the carbon dioxide content of the atmosphere, with consequences that are difficult to predict. In the proposed use of biomass the change of carbon dioxide content in the atmosphere would be compensated. Photosynthesis removes carbon dioxide from the atmosphere and the subsequent burning of the fuel restores the carbon dioxide. All this could be done in a virtually closed system. We might in fact erect an electric generating plant that burns the fuel derived from algae right next to the areas producing the algae. We might then divert the carbon dioxide from the stack, make it available to the algae, and thereby stimulate their growth.

The use of biomass is an attractive idea and possibly quite a feasible

---

[5]T. W. Jeffries, et al., *Biosolar Production of Fuels from Algae* (Lawrence Livermore Laboratory Report No. UCRL-52177, 1976).

one. One may raise the objection that in the not-so-long run the need for food will become even more urgent than the need for energy. Instead of growing biomass for energy, we might rather want to grow food. For instance, vegetation in the water, not digestible by humans, could well be used in fish farms that would deliver badly needed protein.

This proposal does not preclude the use of biomass for energy; it only modifies it. By-products of agricultural output could very well be converted into energy. Corn husks could serve this purpose. An even better example may be bagasse, a residual product of the processing of certain plants such as sugarcane. In Hawaii it is burned and utilized. With bagasse the material is already available in a form where there are no expenditures for collection. In Brazil a large-scale enterprise is underway for the growth of sugar cane for producing alcohol to be used as fuel for automobiles.

Melvin Calvin, in discussing energy through photosynthesis,[6] has suggested another way to use biomass without diminishing the prospective food supply. He reports that some members of the Euphorbia family, which grow in arid regions, produce latex similar to that obtained from rubber plants. This latex lacks the appropriate molecular chains necessary to produce rubber. But after processing, the fluid can be used as fuel or can constitute the raw materials for industries that otherwise would consume oil or gas. The latex of some species of Euphorbia can be gathered year after year by tapping. This suggestion would be particularly valuable if plants could be hybridized to deliver great quantities of latex. Similar feats of biological engineering have been performed in the rubber production of Malaya, increasing the yield per acre tenfold or even more. Dr. Calvin has had encouraging results experimenting with other species. *Euphorbia lathyrus,* for example, an annual presently used in experimental plantings, is cut down and its oil extracted. It is expected to produce about ten barrels of oil per acre of previously nonproductive land, and even its cellulosic residue, bagasse, can be burned as fuel.

Ultimate success in producing fuel from solar energy would come if we could imitate the photosynthesis by which plants convert carbon dioxide, water, and sunlight into sugar or starch. We could thus use the biomass process without the biomass. This would first have to be done in

[6]Melvin Calvin, "Photosynthesis as a Resource for Energy and Materials," *American Scientist, 64:* 270–278 (1976).

FIGURE 12-4 *Fast-growing* Euphorbia lathyris, *shown here in experimental planting in southern California, is expected to produce about ten barrels of oil per acre per year.* E. lathyris *grows without irrigation on previously nonproductive land, reaching a height of about one meter. Mature plants on this "gasoline tree plantation" will be cut and crushed to extract the oil. Even the cellulosic residue, bagasse, can be burned for fuel. (Courtesy of Lawrence Berkeley Laboratory, University of California.)*

the test tube and later in a big-scale industrial process. Unfortunately, the process we wish to imitate is complex. Four light quanta are captured and stored in some intermediate form with the help of the plant's chlorophyll. Indeed, the chemical energy in natural fuels is reached by capturing and adding these several bits of sunlight. We are beginning to understand how plants perform the essential trick. But a long and difficult road will have to be traveled from initial understanding to successful imitation and eventually to economic exploitation in a factory. If this goal could be accomplished we would probably gain deep insight into a basic biological process connected with the development of life on earth, another example of how the intellectual pleasure of understanding and the practical profit from exploiting nature go hand in hand.

Burning the waste products of a city can be considered under the heading of using biomass only in a somewhat unnatural manner. These by-products are, of course, pollutants and must be disposed of at consider-

FIGURE 12-5 *Plants such as this "milkweed,"* Ascelpias, *in Brazil, can serve as hydrocarbon sources. Also found extensively in Puerto Rico, the perennial plant grows in semiarid desert regions to a height of eight feet. (Courtesy of Lawrence Berkeley Laboratory, University of California.)*

able expense. It has been estimated that to convert parts of them into fuel or to burn them in a controlled and useful fashion could supply as much as 2 percent of the energy in the United States. In themselves the waste products will probably not supply an economic source of energy. But if they are not used they have to be disposed of in some fashion. The differential cost between use and disposal will probably turn out to be low enough so that these waste products should actually be made part of our energy supply. Processes for the dual purpose of waste disposal and energy generation have already been initiated in Miami, St. Louis, and a few other places.

The heating of the earth, uneven because of varying geographic latitudes, evaporation of water, and various effects of clouds, drives the great worldwide wind systems. These systems are influenced by topography and are finally dissipated in smaller local wind patterns. Their ultimate driving force comes from the sun, and their total available energy, though much smaller than the energy in sunlight, is still very great. As with sunlight itself, wind energy is difficult to exploit because it is unconcentrated. Having wind is not sufficient to drive a windmill with reasonable efficiency; it is necessary to have strong wind. If wind velocity is doubled the power of a windmill increases eightfold—twice as much air arrives and each volume of air carries fourfold energy. If we want to erect windmills it is essential that we locate them where the wind is strongest.

It is also important to look for steady winds; otherwise storing energy will become an expensive problem. As in the case of solar electricity, this is one of the difficulties to be overcome.

One region in the United States offering particular hope for the use of wind power is Hawaii.[7] The island of Oahu especially, with its million inhabitants, needs energy; today its supply is derived from expensive oil. The Hawaiian Islands lie in the zone of the Trade Winds, which blow 80 percent of the time with a rather steady velocity somewhat over five meters per second (ten knots). This is actually not enough for efficient operation of windmills of familiar design.

But populous Oahu has a rampart of hills on the windward side. As the winds blow over and around the end of these hills, their speed is doubled. The resulting average of ten meters per second suffices for economic windmill operation. It is important to pick just the right spots for these windmills. Since the direction of the Trade Winds is practically fixed, from the east-northeast, the winds are closely related to the topography. Application of computing machines to aerodynamics gives us a reasonably good way to calculate wind patterns and therefore establish the optimal locations. But experimental proof is still required.

The usual way to find the wind pattern is to use anemometers, which are simply three cups carried on a stick, not unlike a child's toy. To measure winds with these instruments at various places and times would obviously be laborious. What is worse, winds are feeble near the surface, where the anemometers can be used; higher winds at an elevation of fifty to a hundred meters should be measured. Indeed, a good modern windmill is a structure on a sizeable tower, having big propellers whose two foils are similar to the wings of an airplane. There is an even greater

[7] In a lecture, when I asked the audience to guess where we have the strongest winds, a facetious answer was "Washington, D.C." An analogous answer in New Zealand is correct. Their parliament is in Wellington, on the shore of the Cooke Strait separating the North and South Islands. The winds, hemmed in between these two land masses, become very strong.

FIGURE 12-6 *Wind on Koko Head in Hawaii. The author is using a handy anemometer to find wind patterns instead of the proper instrument, which is a laser.*

similarity to a fan or an airplane propeller; only in those cases, air motion is produced whereas with windmills the air motion is used. Cause and effect are thus interchanged.

We need a faster, more efficient way to measure winds at various locations, elevations, and times. This may be accomplished by use of laser beams.[8] A laser emits light that can perform almost any job. The beam can be emitted at high intensity, it can be emitted for a very short time

[8] W. M. Porch et al., *Astronomical Techniques Allied to Pollution Detection: Pollution Transport* (Lawrence Livermore Laboratory Report No. UCRL-79484, 1977).

FIGURE 12-7 *Wind turbine at Sandusky, Ohio, which would produce 100 kilowatts if the wind were blowing. (Courtesy of Solar Energy Information Services.)*

interval, and, what is more important, it can be emitted with a sharply defined frequency. The wind itself is invisible, that is, it hardly scatters light. It almost always carries particles, however. In Hawaii it carries small droplets derived from ocean spray. These droplets scatter the laser light. Light scattered from a moving particle will have a different frequency due to the Doppler effect, the effect that causes a siren to be heard at a higher pitch when it is approaching. By measuring the frequency of the scattered light, we can obtain the velocity of the droplets, which is essentially the same as the velocity of the wind that carries the droplets. Furthermore, laser beams can be emitted in very short time intervals and the whole beam may be no longer than one meter, in spite of the tremendous speed of light. By observing the scattered beam at various time delays after emission, one can get information about wind velocities in the location where the scattering process occurred, meter by meter, along the whole line illuminated by the laser.[9] On cloudy days the laser can be supplemented by radar, radiation of longer wavelength that is not stopped by droplets.

The swift action of lasers is one of the reasons why this method of surveying wind patterns will succeed. With one laser, mounted in a truck or possibly an airplane, you can rapidly obtain information about the movement of air as far as you can see. Of course, it is also necessary to observe the scattered light in at least two locations in order to get the direction as well as the velocity of the air movement.

Investigations in Hawaii have progressed to the point where excellent locations for windmills have been found on the northern and western tips of the island of Oahu. Windmills near those locations could deliver one megawatt apiece. In order to supply electricity for the whole population, as many as a thousand windmills would be necessary, and environmentalists have objected to the insulting "visual pollution." I have tried to reassure them by suggesting Christmas decorations on the windmills, but this seemed unacceptable because then the windmills would not look "functional." Fortunately, the northern and western tips of the island do not happen to be famous scenic spots, though all of Hawaii is scenic.[10]

Hawaii is also appropriate for windmills because of its moderate climate. Icing is, of course, no problem. Strong winds are needed, but no hurricanes. There are hurricanes in Hawaii, but the story is that in those fortunate islands even the hurricanes say *"aloha."* The strongest on record produced a wind of only one hundred kilometers (sixty miles) per hour.

In the United States, Hawaii is the best place for windmills; if windmills will not work there they will not be really effective in any other location except some thoroughly isolated spots such as the Aleutian Islands. In the event of success we will have gathered some experience, and may be ready for applications on the mainland. Figure 12-8 shows some

[9]This process is similar to the way in which a bat measures distances. It emits ultrasonic signals and judges distances by the length of time it takes for the echo to return. The bat may be after an insect; we are after droplets carried by the wind. A significant difference is that sound is not all that fast. Light moves a million times faster. The eardrums and nerves of the bat, good receptors for his needs, are enormously surpassed by the apparatus connected with lasers. This is made possible by the miracles of electronics.

[10]In the first half of the last century there were ten thousand windmills in England, which is a tight little island, but still bigger than Oahu. Holland can hardly be imagined without windmills. But these are antique. And so will the windmills we are planning now become if we wait a while.

preferred areas. The figures on the map give the wind power averaged over a year. The shaded areas are mountainous, and spots of particularly high wind velocities can be located there. Before windmills are deployed a careful, year-round study of local winds, preferably one employing lasers, should be carried out.

In Israel there are strong, fairly steady winds in the Arawa Valley, blowing along the big crack in the earth of which the Dead Sea and the Red Sea are parts. They are not as strong or steady as those in Hawaii, but the isolation of the region makes it desirable for establishing the smaller electric generators that can be driven by windmills.

A worldwide survey for the purpose of obtaining favorable locations for windmills would be most helpful in establishing appropriate energy sources at optimal positions.

I have concentrated so far on the wind rather than the windmill. A disadvantage of the present design is that it has a horizontal rotating axis high on a tower. Furthermore, the passage of the foils in front of the tower, or behind it, causes disturbing vibrations.

Among the many and often outlandish novel designs, one ingenious proposal should be mentioned. We could build windmills with a vertical axis that could be more easily connected to an electrical generating unit. These windmills have three twisted foils and the whole structure looks somewhat like a huge upside-down egg beater. They have a further advantage that they do not have to be oriented according to the wind direction, but will accept the wind from any quarter.

The most expensive parts of a windmill are the blades. It makes sense to increase the wind velocity in a smaller area and then use smaller blades to better effect. An ingenious and extreme application of this procedure has been suggested by an engineer from Taiwan, Dr. James T. Yen.[11] He suggests a building consisting of two wings leading into a central cylinder. From the top the outline looks as shown in Figure 12-10a. Furthermore, a horizontal platform divides the structure into lower and upper portions. In the lower portion the cylinder is a dead end so that the wind, being concentrated by the wings, will build up pressure. In the upper portion, veins are added which will cause the wind to arrive tangentially on one side of the cylinder. Thereby a whirlwind will be set up as indicated by the curved arrow in Figure 12-10b.

---

[11] James T. Yen, "Tornado Type Wind Energy System," *IECEC 1975 Record*, p. 987.

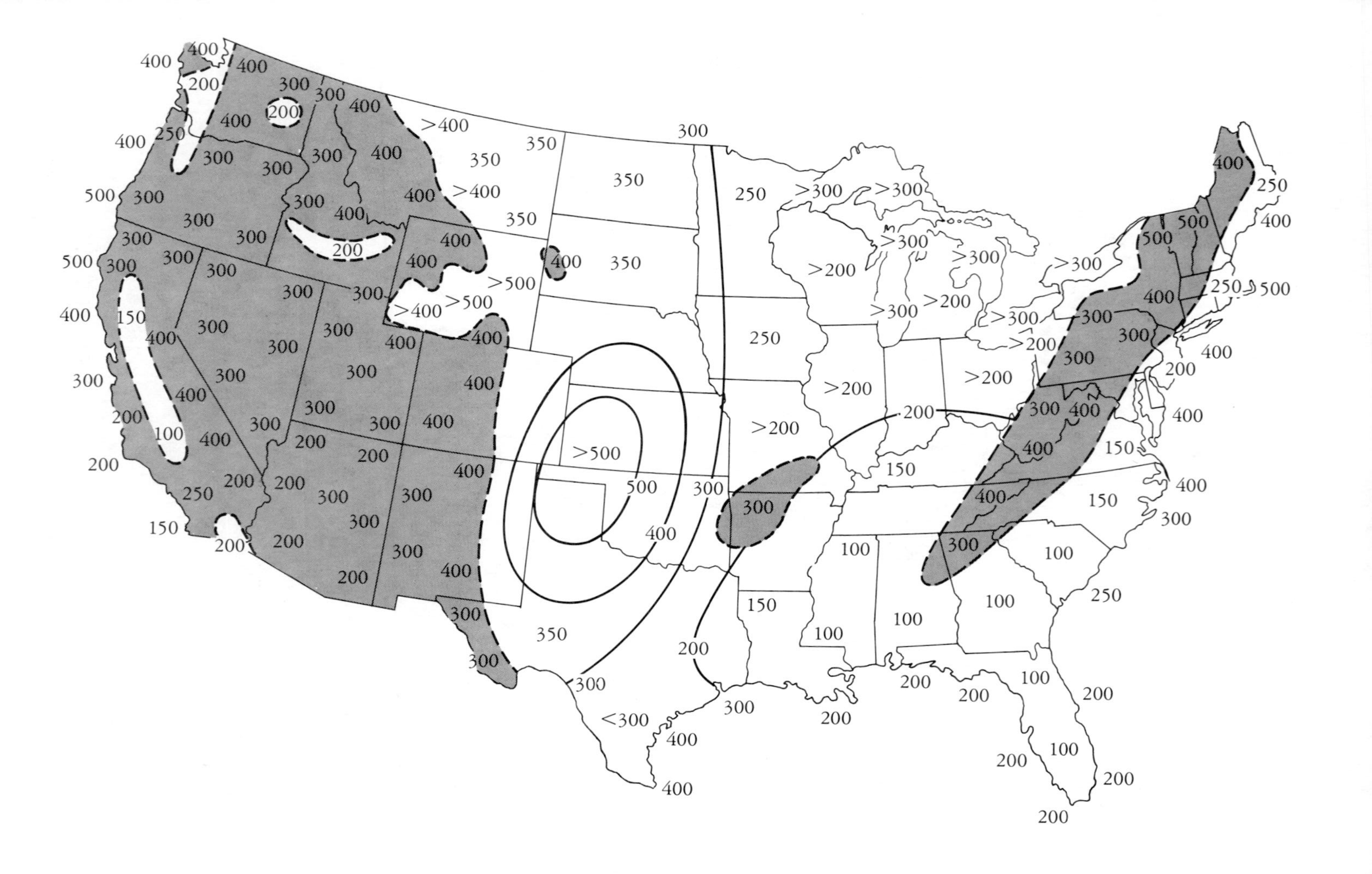

400
400
400
200
300
300
200
400
400
400
>400
300
250
350
350
400
300
300
300
400
250
>300
>300
400
500
300
300
400
>400
500
500
250
400
300
400
350
300
300
>300
300
400
200
400
350
>200
>300
300
500
300
300
>500
>300
250
500
300
300
300
>400
>500
>300
>200
400
400
150
400
300
>300
300
300
400
400
250
>200
300
300
300
>200
>200
200
300
400
>200
300
400
200
400
300
400
100
300
400
>500
400
150
200
200
200
400
150
500
300
300
400
250
200
300
300
150
200
400
300
100
300
400
350
200
300
300
100
100
200
300
200
200
400
<300
400
400
100
200
200
200
200
200

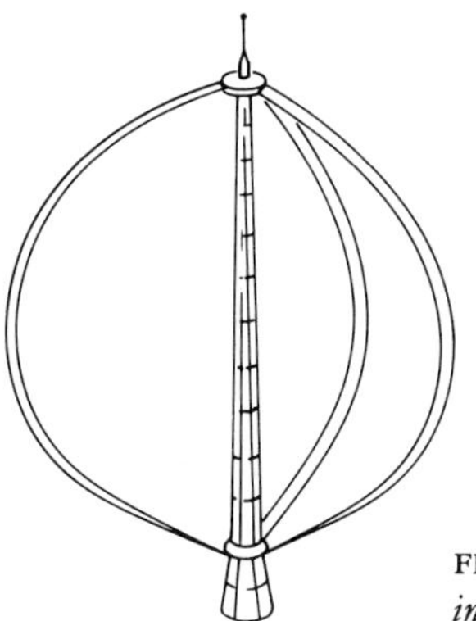

FIGURE 12-9 *Arrangement of foils in a windmill with a vertical axis.*

The upper portion of the cylinder is open at the top so the whirlwind can blow out at the top, forming a small tornado. Due to centrifugal force a region of very low pressure will exist at the center of the tornado. There is an opening in the horizontal division that separates the lower portion from the upper, and that opening is near the center of the cylindrical building. A turbine is placed in the opening, and there will be high pressure below the turbine and low pressure above it. In this way the energy received by the opening of the building's two big wings can be concentrated on this relatively small turbine, which then can deliver energy to an electric generator.

Construction of wings and cylinder would, of course, be quite expensive. It would be best if these structures could be used for an additional

---

FIGURE 12-8 *Mean annual wind power for various areas of the United States. The figures indicate average wind power in watts per square meter, at 50 meters (165 feet) above the surface of the earth. If we wish a windmill to deliver, on the average, 1,000 kilowatts (1 megawatt), its diameter would have to be 100 meters, with the tips of the blades extending back to the earth, which is impractical. The wind power changes rapidly with wind velocity, and therefore the figures are more closely connected with the frequency of suffiiciently high winds than with the average velocity of the wind. The shaded areas are mountainous, and in these regions the average wind power changes rapidly with the location. At selected spots, actual figures for wind power in these regions reach considerably higher values than those given on the map. The sign > 400 means "greater than 400," and the sign < 300 means "less than 300."*

FIGURE 12-10 *A structure that (a) traps the wind inside a cylinder, where (b) a whirlwind is created.*

purpose. One might suggest, for instance, that the whole structure be a hotel, with the most desirable rooms near the tip of the wings and no rooms for guests at the cylinder, since that area is apt to be noisy. The best location for such a hotel would be where the wind is strong and blowing always from the same direction. An ideal location would be on Maui, at the relatively narrow strip of flat land connecting the two big mountains that form this Hawaiian island. If the wings opened toward the Trade Winds, almost steady operation could be expected. Such a hotel, were it ever built, should be called the Maui Tornado Hotel.

The sun drives the winds; the winds drive the waves. In the great oceans these waves carry vast amounts of energy. Far from the shore lines the waves are smooth, and are capable of delivering energy in a relatively simple way. (Long ocean waves carry the most energy; they retain it for days until they hit a shore.) What we need is a steady platform, so that the wave motion relative to the platform can be exploited.[12]

Wave motion lessens as we go deeper. At the depth of a full wave length (the distance from crest to crest), velocities in the water decrease to one four-hundredth of what is observed at the surface. A platform could be mounted on hollow cylinders that go down to a depth where there is practically no wave motion. The platform should be built above the range of the waves, and will be very steady. It could be anchored, or propellers may be used to keep the platform near a prescribed location. These propellers could be driven with comparatively little energy expenditure.

---

[12]Extensive and rewarding research on the utilization of energy from waves, tides, and currents continues, (for instance, at the Scripps Institution of Oceanography, La Jolla, California, and at the Bionics Research Laboratory of the University of Edinburgh, United Kingdom. See Arthur Fisher, "Energy from the Sea," *Popular Science,* 206, No. 5 (1975); and Gerald L. Wick and Walter R. Schmitt, "Prospects for Renewable Energy from the Sea," *MTS Journal,* 11, Nos. 5 and 6 (Marine Technology Society, Inc., New York, 1977).

Hollow iron spheres could be attached, floating, to the platform; they will move up and down with the waves. Their motion relative to the platform could activate pumps to raise water in containers carried by the platform. There are many other, and probably more practical, ways to perform the pumping action. The water should thereafter be allowed to return to the sea through a turbine.

In one area of the world the waves are particularly strong and effective, and exploitation there holds great promise. South of Patagonia and New Zealand, north of the glaciers of the South Pole, are stretches of ocean where one can sail around the earth without ever bumping into land. The waves do not dissipate their energy by pounding on shorelines. They grow to an average height of ten meters from trough to crest. That is where the best wave machines might be floated.

Generation of electricity by this platform method would be straightforward; its exploitation is more difficult. Transmitting the electrical energy through long cables to a land mass would be expensive. But how much electricity would be needed in Patagonia or Tasmania? Perhaps a little more would be useful to southernmost New Zealand, a country with an excellent electric network. It might be better to use the energy right on the platform. There are many ways to produce easily transported goods. One possibility, though almost certainly not the best one, is connected with the exploitation of manganese nodules lying on the ocean floor, spheres with a high content of valuable metals such as copper, nickel, and manganese. These nodules could be refined on the platform itself, particularly if the platform were located in a region rich in the nodules. According to present technology, electrical energy offers one possible way of refining the nodules, though not the best one.

Another possibility would be to produce aluminum from bauxite, which is rather pure aluminum oxide. That is a process giving a valuable product, which requires electricity. Shipping the bauxite from its usual place of origin, the wet tropics, should present no serious problem.

The idea of large, stable platforms at sea has been given a little consideration. One possible function, and this is indeed an ambitious proposal, would be that a small city could be built on such a platform. To obtain a total area of a square mile for urban use at a cost of a billion dollars may not be an unreasonable estimate. For one hundred thousand inhabitants this would amount to $10,000 per person in initial investment. To move the platform at a slow three to five knots would be relatively easy. Such a city would have the advantage that it could locate itself in a good climate all the year round. The main question would be, what is a good climate? The ocean itself, with its many attractive and useful properties, would be an added inducement to potential citizens. The city's energy, of course, could come from the waves. For energy, this would be a case of Mohammad going to the mountain, since, according to tradition, the alternative is harder to arrange.

A popular example of how the sun could provide useful energy is Ocean Thermal Energy Conversion (OTEC). It has been mentioned several times that energy present at a fixed temperature cannot in itself be used; a temperature difference is needed. Such a difference exists in the tropics between the warm surface water and the much colder water at great depth. Several methods have been suggested for using this temperature difference to generate electricity. The most straightforward would be to evaporate liquid by bringing it into contact with the warm water, and to condense it by cooling it with the cold water. In this way an engine similar to the steam engine could be constructed, in principle.

The trouble is low efficiency. According to the laws of thermodynamics, at most only 7 percent of the energy could be utilized, and at least half of that would be lost in actual engineering execution. Such low efficiencies are prohibitive, not because too little energy is left over, since we started with plenty of energy, but because one must operate with volumes and masses of materials that are too great for the extraction of a given amount of energy. The result is high capital investment and eventually high cost of electricity. This statement is not provable and absolute. It only indicates that the great amounts of heat stored in the oceans can be effectively turned into electricity only through truly ingenious ideas.

Solar power is the most popular among novel sources of energy. Indeed, we have seen so many approaches to the use of solar energy that we can be sure that some of these approaches will become practical. Arbitrary estimates of solar energy projections for the United States are shown in Table 12-2 in quads (or 10^{15} Btu) per year.[13]

From our discussion of solar energy, let us turn to energy from the moon. Few people think of the moon as a source of energy for anything

[13] Present energy consumption in the United States is 75 quads; in the world it is 300 quads.

TABLE 12-2 *United States Solar Energy Estimates*

Application	*1982*	*2000*
Heating and cooling	0.2	3
Agriculture	0.0	1
Industry	0.1	1
Biomass	0.2	2
Solar electricity	0.0	1
Wind and waves	0.1	3
Burning of wastes	0.0	2
Total	0.6	13

other than romance. There is another potential application of lunar energy, however. The relative motion of the moon and the earth causes tides, and tidal power stations have been proposed and even constructed. Unfortunately, the energy is not concentrated enough and we again run into the difficulty of high capital cost for a given amount of energy.

There are some specific locations where nature manages to concentrate tidal energy. One of these is the Bay of Fundy, which separates most of Nova Scotia from mainland Canada. In a nearly enclosed body a period of oscillation occurs for the water that is crudely analogous to the period of oscillation of a pendulum. In the Bay of Fundy this period is approximately twelve hours, which also happens to be the period of the tides. Because of this "resonance" the tides are reinforced, and in the Bay of Fundy high tide and low tide differ by twenty meters.

This tidal motion could be exploited in one relatively simple way. Structures similar to windmills placed in the bay could be activated by the water streaming into the bay and out again. The total energy that would be available is several thousand megawatts. That amount corresponds to 1 percent of the electric power in North America. In this special case practical execution is a possibility, particularly if local use for this big amount of energy could be found, or alternatively if energy transmission proved not too expensive.

A somewhat similar possibility exists in exploitation of the Gulf Stream. (Since the water has been heated by the sun and driven, in the ultimate analysis, by the pattern of the winds, this really should be considered as solar energy.) There are regions in the Gulf Stream where the water velocity is well over three knots and where a possibility exists for effective conversion into electricity. Because of the greater density of water, three knots in water compares with a windmill driven by wind of about twenty-seven knots. There are, however, two strong arguments against this utilization. One is that the Gulf Stream could account for considerably less than the electrical requirements on the east coast of the United States. The other point, much more important and worrisome, is that interference with the Gulf Stream could have a considerable influence on the climate of western Europe. There seems to be reason to suspect that it is the Gulf Stream that prevents France and England from having the climate of Labrador. This is, therefore, an area where worry about environmental effects could be amply justified. Possible harmful effects illustrate global interdependence in the modern world.

Among the general uses of solar energy as it is delivered daily to the earth, we must include hydroelectric power. The evaporative action of sunshine creates the clouds, the eventual rain, and the resulting energy available in the rivers. It is worth remembering that use of water power was one of the earliest indications of the coming Industrial Revolution. Another indication was the proliferation of windmills. Neither the water wheel nor the windmill was invented in medieval western Europe. The Greco-Roman civilization used water wheels; windmills came from Persia. Both, however, became popular throughout Europe during the Middle Ages.[14]

At present, big masses of water behind huge and impressive dams provide a considerable portion of electric power. In the early days of electricity, water power was almost its only source. Today in the United States most of the available hydroelectric power has been developed. Great additional amounts could be made available only in Alaska, on the Yukon River, for instance. Unfortunately, a big Yukon hydroelectric project that could become useful at some future time would create a lake that would

---

[14]Perhaps it is not quite appropriate to discuss water power in this chapter. The technology is not new and has practically none of the properties of a new technology. Additionally, it is not necessarily tied to solar power. Water reservoirs work much the same whether they are filled by rainfall or by water pumped out of a river by some other source of power so that electricity may be stored for the hours of peak demand. But among my priorities, a systematic exposition is not high. Systematics may be left to the dictionaries.

inundate the Alaskan pipeline. Throughout the world, however, hydroelectric power still could provide significant additional amounts of electricity. Central Africa, which is especially low in per-capita energy consumption, is a favorable prospect for water power.

It should be clear, nonetheless, that hydroelectric power has some serious disadvantages. It is dangerous and, contrary to general opinion, it is in a practical sense not renewable.

As mentioned in Chapter 10, dams have collapsed due to earthquakes and other causes. Thousands of people have perished in such disasters. The construction of dams has developed slowly together with engineering practice. Scientific foundations for the prediction of trouble may not be as clear-cut as in the example of nuclear reactor technology, developed in a period where everyone was acutely aware of the dangers. It should be remembered that big hydroelectric dams bear the great burden of water masses. As the water infiltrates into cracks at the bottom of the lake formed behind the dam, small tremors are observed. We learned not many years ago that water can lubricate rock masses and make displacement easier. How these phenomena may be connected with the development of earthquakes is unclear.

Deployment of gauges that continue to measure strains in dams as they change with time may be used for better understanding and eventually for greater safety. Actual small deformation of dams should also be continuously observed. Finally, the developing art of earthquake prediction may help to improve the safety of dams. I have visited huge dams in an earthquake-prone island, Java. I was disturbed about the safety outlook. Yet energy is needed in Java. We can try to understand, try to help them improve, and pray.

The energy source of hydroelectric power plants is of course continuously renewed as the artificial lake fills up in the rainy season, providing power the year round. But these reservoirs will silt up in a century. So far as we now see, it would then be necessary to shift to another site. Not only is resiting expensive, but new favorable sites are apt to become nonexistent. Therefore, it is yet to be proved that hydroelectric power is a truly renewable resource. It may become so if technology should find the proper answers to the difficulties mentioned.

Another energy source that many people call "renewable" is the heat contained in the interior of the earth. Actually, there is a heat flux from the interior of our globe that is as old as the earth, five billion years, but it is insignificant. The significant geothermal sources are in a way comparable to fossil energy. Coal has been deposited in the last few hundred million years from plant life. Hot local lava deposits have been established in the last few million years in the upper few miles of the earth's crust. The former, coal, has been used extensively. The latter, the useful form of geothermal energy, has also been used, but so far only in a few instances. Outstanding examples are geothermal works in Italy, New Zealand, Iceland, and most particularly near San Francisco in the Geyser project. This last-named project supplies electricity for San Francisco, and is indeed one of the least costly of generating plants.

One further similarity between fossil fuels and geothermal heat is that total usable amounts of energy are comparable. Geothermal heat should therefore be considered as an important source of energy, but no more renewable than coal. Indeed, we can call it fossil volcanic heat.

There are particular regions, of course, where geothermal energy should be sought. The western United States is one such region. It is part of a much bigger region: the whole rim of the Pacific Ocean. Modern geology has developed the theory that moving plates exist in the upper layers of our globe. These plates appear to move slowly but steadily, as more or less rigid entities. Seismic disturbances, that is earthquakes and volcanoes, frequently occur on the boundary between plates where the plates move in relation to each other. The slow motion is sufficient to bring about the tensions and disturbances that characterize seismically active regions.

The rim of the Pacific, sometimes also called a rim of fire, is a collection of such active regions. An excellent geothermal field within this rim is Wairaki, which lies in the middle of the North Island of New Zealand. It is a relatively small part of an extended field stretching to the northeastern coast of the island. This rich field is the best candidate for cheap energy in New Zealand. Java, Japan, and the big island of Hawaii are some other promising locations that so far have been less well explored and defined.

The most easily developed geothermal energy sources are regions where water is in contact with very hot rock due to prehistoric lava flows originating in the deeper interior. The water itself frequently comes from rainfall that has trickled through the layers. This water is often called meteoric water, "meteoric" meaning that, like the meteors, it has come from the sky. The water gets vaporized and can give rise to hot springs, geysers, and the like.

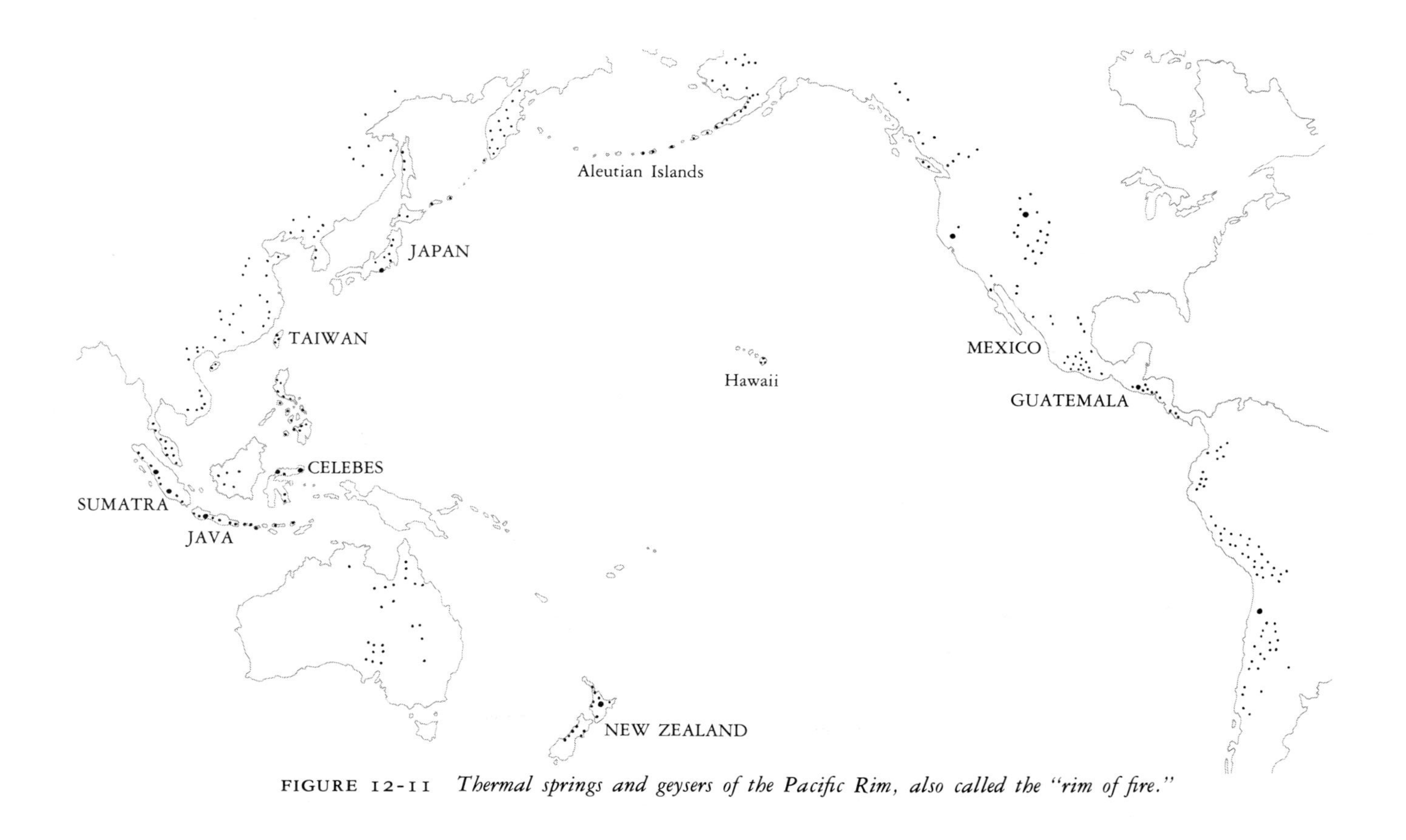

FIGURE 12-11 *Thermal springs and geysers of the Pacific Rim, also called the "rim of fire."*

High-quality geothermal sources produce rather pure vapor with few contaminants, through contact with high-temperature rock. The excellent source near San Francisco is almost pure, although it contains slight amounts of hydrogen sulfide, which has a disagreeable odor and, when present in bigger amounts, is poisonous. Geothermal energy does, in general, cause some pollution.

Unfortunately, high temperature and relatively clean geothermal sources, such as those near San Francisco or in New Zealand, are the exceptions rather than the rule. Another massive and rather clean geothermal source is available in Yellowstone National Park; it is the biggest in the United States. It resulted from a great volcanic eruption that occurred less than a million years ago. It could be used easily except for two circumstances. The demand for electricity in that region is somewhat limited, with Denver and Salt Lake City the major consumers. The other objection is that Yellowstone, a national park, is not supposed to be disturbed. Actually, exploitation of Yellowstone Park's geothermal potential would hardly disfigure the landscape. In the long run, however, Old Faithful might not work quite as predictably.

We have not yet gotten to the point where we are willing to sacrifice a tourist attraction to our economic needs. It is unclear whether this is due to our unwillingness to accept change or primarily to a conviction that electricity will always be available when we turn the switch, whereas we ought to preserve the wonders of nature. To prefer the interest of tourists to daily necessities is a modern and peculiar American brand of a deep truth: beauty is more important than comfort.[15]

It is difficult to make a choice between our needs and our value judgments concerning natural beauty. Today most vocal people will strongly advocate beauty rather than comfort. It will be necessary to consider carefully what kind of beauty we must preserve, and to determine what comfort may be a necessity for our way of life.

In the special case of Yellowstone Park the answer at this time will almost certainly favor unexploited nature. In that region of sparse popula-

---

[15] A great physicist, Werner Heisenberg, has defined a deep truth as one of which the opposite is also a deep truth. In our case we might tend to formulate the opposite with a slight variation: need is more important than beauty. A brief statement of wisdom from a fortune cookie in a Chinese restaurant may be used by both sides of the dispute: "Our necessities are few but our wants are endless."

tion it probably is not worth sacrificing any part of the oldest and most famous of our national parks.

Near San Francisco, in Yellowstone Park, and in a few other places, relatively clean steam is available. There are many more places where lower quality geothermal energy can be obtained. Partly the temperature is lower, and if the pressure is relieved, what might be obtained is not good dry steam but wet steam containing a myriad of water droplets. The steam also carries great masses of contaminants, not only hydrogen sulfide but silt and all kinds of salts. The use of such geothermal sources will result in considerable pollution unless we end up pumping the condensed water and salts back under the ground. This process, of course, would add to the expense. Another problem is that the dirty mixture of steam, water, silt, and salt will corrode and erode turbine blades, heat exchangers, and other parts of the apparatus that is to utilize the energy. The salt and silt are also apt to build up deposits in pipes, closing them off. But these difficulties may be overcome. Adding minute amounts of acid to the hot mixture seems to be of considerable help. We also are trying to find coatings to protect the machinery from the adverse effects of the fluids.

The most hopeful approach seems to be the *total flow concept.*[16] Underground the geothermal resource is a liquid because it is under pressure. When the liquid is discharged through a well, pressure is released and the solution partially vaporizes. As the vapor expands, it accelerates, gains kinetic energy, and imparts kinetic energy to unvaporized droplets. In the total flow procedure this whole ejected mixture will fall on the propeller blades of a turbine, which then generates electricity. The approach is simple, requiring no separation of steam, water, and solids, and no delicate heat exchangers. Total flow has yielded an efficiency of 18 percent for the plentiful deposits near the Salton Sea in Southern California, extensive deposits with salinity exceeding that of the ocean. Considering the relatively low temperature of geothermal deposits, this is good efficiency. If all the obstacles to utilization of geothermal sources are overcome, these deposits could satisfy the electric needs of Southern California for several decades.

Another approach to geothermal energy would be to utilize hot rock

---

[16] A. L. Austin et al., *The Total Flow Concept for Recovery of Energy from Geothermal Hot Brine Deposits* (Lawrence Livermore Laboratory Report No. UCRL-51366, 1973).

that does not contain water.[17] In this case water would have to be pumped down. The problem is to make sufficient contact between the water and the rock. This in turn means that very extensive cracks would have to be created in the rock, a procedure that seems difficult. In efforts to increase the permeability of rock containing natural gas, the creation of such cracks by pumping in water or by means of explosives has been thoroughly explored. Success has been marginal at best. Yet the energy content of natural gas in good gas-bearing deposits is approximately a hundred times as valuable as that of hot rock of the same volume. Therefore, the economic success of exploiting dry hot rock would seem doubtful. To create surfaces in hot volcanic rock is, of course, quite a different problem from creating added surfaces in a gas-containing sedimentary rock formation already full of microcavities.

Geothermal energy may bring along its own peculiar danger. Nuclear reactors or hydroelectric dams may suffer from earthquakes. Use of geothermal heat may bring about earthquakes. It has been found that water acts as a lubricant on hot silicate rocks. In the virtually earthquake-free region near Denver, when water carrying waste materials was pumped into deep wells, earthquakes of moderate size resulted. Futher experimentation established a clear cause-and-effect relationship. Of course, the earthquake results from preexisting tensions. But the water serves to release them, a circumstance that should be kept in mind, particularly in "dry" geothermal procedures, where water is pumped down into layers that had previously contained none.

Another energy resource related to geothermal energy is found along the coastline of Texas.[18] Here, deep underground, great amounts of moderately hot water are subjected to the pressure of the rock formation that lies above them. This pressure of rocks is called *lithostatic pressure,* in contrast to hydrostatic pressure, which is pressure exerted by the weight of water. A well drilled at the shoreline would cause water to squirt up with a kinetic energy due to the rock pressure. An interesting added fact is that the huge amounts of water contain a small percentage of natural gas, which, considering the great volume of water, amounts to significant

---

[17]M. C. Smith, *Energy and Mineral Recovery* (U.S. Department of Energy Document No. CONF-770440, 1977), p. 54.

[18]Myron Dorfman et al., *Energy and Mineral Recovery* (U.S. Department of Energy Document No. CONF-770440, 1977), p. 562.

quantities. Part of this natural gas is released as pressure is reduced. Three energy sources—the kinetic energy of the water, its heat, and the natural gas released—add up to a resource which may be exploited with economic advantage. An environmental difficulty would be subsidence of the area where the water has been allowed to escape.

Geothermal energy has been exploited in only the few cases where use is most straightforward. Coal was used for many centuries before it became a major economic commodity. It is obvious that uses of geothermal energy can be expanded, but to what extent, and at what cost, remain unknown.

Solar energy and geothermal energy are the two major potential sources of new energy. Additional improvements that may gain importance are not really new resources but new procedures for the use of resources.

Let us consider the hydrogen economy, by which we mean a cycle whereby we use energy to release hydrogen from water, regain the energy by burning the hydrogen, and arrive back at water.[19] Hydrogen is an excellent fuel, and from the point of view of energy per unit of mass it is the best chemical source, though it is inferior to gasoline if you consider energy per unit of volume.

Offhand, the burning of hydrogen would seem to be pollution-free. When combined with oxygen it produces water, and of course there are no ashes. Unfortunately, in the burning process oxygen and nitrogen molecules are partially decomposed into atoms; then they reunite. But in some cases they don't find the right partner, and instead of nitrogen and oxygen molecules (N_2 and O_2) they produce nitrogen oxide (NO). This substance has dangerous derivatives, some of them carcinogenic (they have been mentioned in Chapter 7). To call the burning of hydrogen completely pollution-free is therefore an oversimplification.

Hydrogen, as a general rule, is not a natural resource. It occurs as a minor component in natural gas. If it should be used extensively it will have to consume other energy sources, whether through electrolysis or by appropriate chemical processes, for instance, those connected with coal gasification.

---

[19]See, for example, J. H. Kelley and E. A. Lanman, *Hydrogen Tomorrow* (Pasadena: Jet Propulsion Laboratory Report No. 5040–1, 1975).

In a hydrogen economy the hydrogen itself essentially represents a form in which energy has been stored. One of its additional advantages is that it is particularly easily transmitted through pipelines. It also has several disadvantages. In liquid form it is highly flammable and is more dangerous than liquified natural gas, although it is harmless when handled with great care. It must be liquified if it is to be transported in ships, and it will be necessary to make the ships comparable in size to oil tankers and liquid natural gas carriers, in order to limit the cost. From the point of view of possible accidents, this transportation method would have to be scrutinized more carefully than any other energy source.

If hydrogen energy ever should become feasible, the first application may well be the fueling of the biggest aircraft, weighing approximately five hundred metric tons. In airplane application the light weight of hydrogen is of particular importance. The relatively big volume of liquid hydrogen per unit of energy content is a severe disadvantage for small planes, but becomes less important for bigger ones. A crash would, of course, result in a disastrous fire, but in the case of gasoline-fueled planes the fire hazard is almost as great.

In any case, a hydrogen economy would mean extensive work at low temperatures, which in turn would mean large-scale application of new and somewhat hazardous techniques. It is unclear whether or how soon the advantages of hydrogen will justify the considerable technical effort it would involve.

Hydrogen could be used in propelling automobiles or vehicles of the future resembling automobiles. In automobiles, hydrogen might not be used in liquid form, but rather dissolved in solids. In this form it would hardly be dangerous, but the weight of the solid would be a disadvantage.

Substitutes will be needed as oil is depleted, and one such substitute is hydrogen. Others are oil-like substances that can be obtained from coal liquefaction or from oil shale. However, one should not exclude the use of other combustible chemicals for which there are as yet no good concrete suggestions.

Another form of utilizing energy is the fuel cell, which has been discussed for a century. In principle the fuel cell is a battery that produces an electric current, not by the chemical process of oxidizing lead or using some other solid chemical reaction, but by the reaction between hydrogen and oxygen, or even the reaction between petroleum and oxygen. The essential point is that the burning of gasoline and oxygen is utilized in existing motors with an efficiency not exceeding 15 percent in an internal combustion engine, or 40 percent in a carefully constructed utility plant for generating electricity. On the other hand, a fuel cell can operate at nearly 100 percent efficiency with very little heat being wasted. This is true in the ideal, or "reversible," case in which the cell can be run either forward, producing electrical energy from fuel, or backward, producing fuel from electrical energy.

Such fuel cells exist. Utilizing the hydrogen-oxygen reaction, they have been employed in our space program, where expense was less important and minimum weight most important. Existing fuel cells are unfortunately quite expensive. For instance, to make hydrogen available to the electrolyte in which the moving ions produce electricity, hydrogen is dissolved in expensive platinum. How to make fuel cells cheaply, and how to use readily available materials such as gasoline in them, are questions that have tantalized electrochemists for generations.[20]

The last couple of topics mentioned are really novel procedures for storing or using energy. A matter of equal importance is how to transmit electricity over long distances with minimum loss and at minimum cost. There has been steady progress in using higher voltage, and also in the application of direct current rather than alternating current.[21] The former has advantages when distances become longer.[22]

We could make really decisive progress by using low temperatures where the resistance of metals becomes small. Thus, more current could be carried and losses would be decreased. At quite low temperatures, resistance actually vanishes in some metals and semiconductors. This is the remarkable phenomenon of superconductivity.

Low temperatures could be best maintained in underground cables. Of course, locating cables underground would also be desirable for the protection of the transmission lines and to avoid unsightly towers, which

---

[20]See Chapter 5, pages 91–92, where fuel cells are also discussed.

[21]One limitation to the use of alternating current is that when we try to transmit electricity across the continent, complications arise. In spite of the high velocity of transmission, a sixty-cycle alternating current begins to get out of phase, and the system becomes correspondingly more complex.

[22]In the early days of electricity there was a bitter fight between advocates of direct current, led by Edison, and advocates of alternating current, led by George Westinghouse. Alternating current won because of economics of transmission at high voltage and performing the transformation between high voltage and low. Proponents of direct current resorted to scare propaganda, similar to current scare stories about nuclear reactors, about alternating high currents.

are quite conspicuous in our landscape. Low-temperature technology, however, is still in the development stage. It is unclear when and whether superconducting systems will become cheap and reliable enough for extensive use.

Most of the prospects mentioned in this chapter are uncertain. Each person evaluating them today is likely to come to a different conclusion. But the greatest shortcoming probably lies not in what has been said, but in what has not been mentioned. To foresee the technology for even a few decades requires imagination and power of judgment equal to the sum total of the best available in science and technology in this generation. To predict the future is difficult. To predict the technological future is impossible.

PLANS AND POLICIES

ENERGY POLICIES

In which the reader finds himself liberated from scientific and technical details only to be enmeshed in general problems that are no less controversial or complex. Besides demands for a clean environment and the need to invigorate research, we face the impossible task of deciding what we do next. We may assume that if we disregard the energy problem it will go away; or we may decide that we have less and consequently we should use less; or we may try to create what we lack; or we may finally regard the problem as a worldwide crisis. In a desperate attempt at impartiality the author fabricates economic arguments that the cost will be approximately the same whichever course we choose.

There have been justified complaints that the United States does not have an energy policy. It is not clear how many other nations have an energy policy. Perhaps the one government with a clear-cut and highly successful policy is the royal family of Saudi Arabia: let's squeeze as much revenue out of oil as possible, but not so much that it brings about the kind of severe economic hardship that will destroy the demand for oil and kill the goose that lays the golden egg. The sheiks have gone to Harvard Business School and have learned their lessons well.

For the United States a policy is needed that would help to eliminate our dangerous dependence on foreign oil. Advocating a policy is not easy. Instead, I shall attempt to discuss several possibilities.

Of course, more than one policy is worse than useless if the policies contradict each other. But policies are needed in more than one respect, and also where a single policy is to be selected, it is better to consider alternatives. Then one may select among them or find a compromise.

As a first step, two energy-related policies should be discussed: a policy for protection of the environment and a policy for research and development. Then I should like to discuss four alternative policies for energy. One is to stick to business as usual, including and perhaps expanding the regulations that have grown up around energy. The second is to minimize our fuel consumption. The third is to encourage the growth of energy usage, helped along by reestablishing the traditional rapid growth of American economics. The last is to consider the American problem within the framework of the worldwide energy shortage. These brief designations will become clear as each of them is discussed and argued.

We turn first to the formulation of policy on energy as it interrelates with the environment.[1] In the last ten years problems of environmental pollution, previously neglected, have become generally recognized public concerns. Branches of the environmentalist movement have actually become quite extreme. One example is that in the face of an impending energy shortage, the pipeline through Alaska was opposed on trivial grounds. Exaggerated environmentalist concerns were rejected by the Alaskans themselves.[2] It would be a real tragedy if exaggeration of this

[1] Some of the following considerations have been touched upon in earlier chapters, for instance in Chapters 5, 7, and 11. These are repeated here from the point of view of environmental concern, in order to give a well-rounded picture.

[2] At the height of the controversy bumper stickers appeared in Alaska: "Sierra Club Go Home!"

type resulted in a reaction and a return to neglect of the needs of our environment. A policy is necessary that will protect the environment in a reasonable way, but that will not unjustifiably interfere with life's necessities, among which energy is particularly important. It should be beyond question that we want a balance. The real question is how to attain it.

I want to argue against one proposal for establishing such a balance, which seems to have the advantage of being simple and self-enforcing. It is the economic policy to "internalize" the cost of pollution. It is simple because it converts environmental damage into dollars, thereby establishing a direct comparison.

The trouble with this policy is that its simplicity is more apparent than real. Money is a convenient and, at least in my opinion, a proper tool for comparison as long as money changes hands according to the individual decisions of many people.[3] In the case of environmental damage the monetary equivalent is never established by individual actions of many people. It is instead established by a bureaucracy. Therefore, it is simpler and more straightforward to acknowledge from the outset that protecting the environment is and must be a function of government that may in some cases be expressed through financial means, and in still other cases—such as effective sewage disposal, for example—through positive government actions. Once the role of government is identified it becomes obvious that we are dealing with a complex problem that will not be solved by a policy which can be formulated in simple, general, yet effective terms.

I would like to suggest a few points which, consistently applied, could lead to a reasonable and balanced environmental policy.

First, we should consider environmental policy in the narrow sense. Some considerations seem obvious, though they have not been followed with any consistency in the past.

☐ A distinction should be made between pollution that damages health and pollution whose only adverse consequences are aesthetic. A simple example is the emission from chimneys and stacks. Part of the emission

---

[3]One must acknowledge that the concept of money as a measure of values is in any case an oversimplification. This becomes obvious if one thinks of the complexities when individual decisions are influenced by government regulations on one hand, and by advertising on the other.

consists of visible, relatively large particles. They are offensive and are regarded as dirt, but do not constitute a direct health hazard. The small-sized invisible particles that can lodge in the lungs are aesthetically acceptable, but can be truly hazardous. It seems clear that elimination of health hazards must have first priority.

☐ As environmental protection became a political slogan, environmental regulations were introduced without thorough evaluation of the potential hazard or side effects from the regulations themselves. For instance, methods adopted to clean up auto emissions did get rid of unburned hydrocarbons, but they also increased carbon monoxide in the emission. All environmental regulations should be thoroughly evaluated for possible hazards and their effects analyzed before adoption.

☐ Where environmental damage can be repaired at a reasonable cost, outright prohibition should be avoided; instead, the spending of funds to repair the damage should be required. An important case in point is surface mining. At many sites, particularly in the western United States, the mined area can be returned to excellent condition for a small fraction of the value of the mined mineral.

The environmental movement has emphasized the need to avoid accidents, but has done so by publicizing and often exaggerating one particular type of accident. In order to establish a balanced view, the following guidelines are recommended:

☐ Comparisons should be made of the hazards resulting from each of the various ways to accomplish a purpose. In our discussion of nuclear reactors it was pointed out that they produce electricity more safely than hydroelectric dams or a number of other sources of electricity. The number of people killed and hurt in automobile accidents is being compared appropriately with the death and injury rate of rail, air, or ship transportation. The comparison between autos and horse-drawn carriages is almost never made, and it is unclear what the result of such a comparison would be. (Of course, the present density of traffic simply could not be obtained with horse-drawn carriages.)

☐ Inquiry into a certain type of accident should bear a reasonable relationship to the probability of that accident. It is remarkable that the effort spent on studies devoted to the danger of liquid natural gas explosions has been less than 1 percent of the effort expended on nuclear reactors, despite the existence of extensive studies showing nuclear reactors to be less dangerous. In particular, mental pictures of the mere possibility of a horrible accident are often given great emphasis without regard for the improbability of the occurrence—an improbability which, if expressed somewhat more loosely, might even be described as an impossibility.

☐ Translating into dollars the danger to human life or health is improper; it amounts to assigning a certain value to human life. Once the point is raised in this form the great majority of people find the procedure not only impractical but repulsive. It is straightforward and entirely possible to compare danger with danger, but not danger with dollars.

☐ The obvious fact should be borne in mind that absence of energy also results in loss of health and is in itself a danger.

One may express the present viewpoint of some environmentalists by the statement that the main environmental pollutant is man himself.[4] The following suggestions might be helpful in this regard.

☐ If a new development, strip mining in Wyoming, for instance, requires a massive influx of people into a sparsely populated area, the industry should provide and the state or federal government should regulate accommodations that would minimize the disorder and pollution usually linked to big population movements.

☐ The aesthetic judgment that undisturbed nature is the most beautiful has, at best, temporary validity. The word "desert" has a negative connotation and implies "deserted by man." I have in my office a color photo showing a steel bridge spanning the Rio Grande, with a mesa in the background and a tree in its autumn foliage in the foreground. I am one who considers the contrast beautiful. It is not the job of environmental regulators to lay down the law on questions of beauty.[5]

Of course, these considerations do not add up to a policy. They indicate that a reasonable environmental policy could be composed of many statements of the kind that, rightly or wrongly, have been proposed above. The best energy policy can be rendered useless by exaggerated environmental requirements. The converse of this statement is equally true. Even an excellent energy supply that encourages industry and helps the economy will leave us dissatisfied if, as a consequence, our environment is thoroughly polluted.

Equally important in formulating an energy policy is emphasis on research and development. Both areas, and particularly research, have

---

[4]In order to escape criticism from the women, I have to admit that women are at least as dangerous.

[5]I have heard environmentalists complain that a wilderness area is being overrun by backpackers. Can beauty exist if it lies in the eye of no human beholder?

positive connotations. Historically in the United States progress has had the greatest prestige. During the last couple of decades progress has lost its glamor, and therefore development leading to material progress is no longer above criticism. Policy and public sentiment are closely related. One vital question is whether research and development should have high priority. This question relates to research and development in energy but has, of course, much wider application. It is my opinion that progress was venerated in the past in an exaggerated, perhaps even blind, manner. The results of this progress have contributed to what the United States is today. Self-criticism is good. The criticism of exclusive emphasis on progress is justifiable. But both self-criticism and criticism of progress can be overdone and are actually overemphasized today.

Energy is a case in point. We face a national problem and a worldwide problem. Without research and development and the progress that comes from them, the energy question cannot be solved. This much should be generally recognized.

But a research and development policy runs into some intricate special problems. These stem from the circumstance that applying research and development is a high-risk undertaking. When it pays off it can pay off a thousandfold. But in most cases it does not pay off. Research is much more a game of chance than is wild-catting.

Another difficulty is the time scale in which research and development produce results. The time factor has led to a rather clear-cut distinction. Industry usually sponsors research that will pay off within three years; the responsible officials want to show a profit to their stockholders within a rather short time interval. Government, on the other hand, usually concentrates on enterprises that take longer. The moon landing was planned and executed in one decade, and as government projects go, this is not a long time span. The work on nuclear reactors in national laboratories was planned twenty years ahead.

Between private enterprise and government action there remains a research and development no-man's-land. Little emphasis is given to research that will show results within five to ten years.

Of course, exceptions exist, and the exceptions prove the rule.[6] The

[6]The meaning of this common saying is a little ambiguous due to the several definitions that the dictionary gives for the word "prove." I like to interpret the statement not in the absurd sense that an exception makes a truth more certain, but in the sense, according to definition number 1 in Webster's New International Dictionary, "to prove means to try or to ascertain," that an exception puts a rule to the test and thereby helps to clarify it.

electronics industry has supported experiments that reach ten years into the future. As a consequence, this industry has thrived. During World War II the United States supported a short-term development of artificial rubber; this was vital in meeting the challenge of that critical period.

The energy shortage appears to be a special case with a need for research and development aimed at results five to ten years hence. Today neither private enterprise nor government laboratories are paying sufficient attention to this vital time span.

Evaluation of research and development, and decisions based on this evaluation, become particularly important where the risk is great and judgment can be rendered only after some delay. These are the twin dangers: either overemphasizing a unique remedy, or pursuing willy-nilly every conceivable line of research. The consequences of poor decision making are obvious. We have committed many mistakes. For example, we have limited ourselves to traditional lines of development, such as erecting pilot plans for World War II-type coal gasification, and coal liquefaction yielding oil inefficiently at $30 per barrel. We have pursued popular concepts such as solar energy without sufficiently defining how this research should pay off within a reasonable time interval. I am tempted to say we have become "experts" in research and development. But if I repeat the definition of an expert given by Niels Bohr—"An expert is a person who through his own painful experience has found out all the mistakes which can be committed in a very narrow field"—I must admit that Bohr's definition does not apply here. Energy research is not a narrow field and the supply of mistakes seems inexhaustible.

We are brought back to the main policy question with all these uncertainties in mind: Where should the research be carried out? Who should make the decisions? How should the decisions be made?

In most contemporary technical literature, science and technology appear as Siamese twins. This relationship corresponds to a phenomenon typical of the twentieth century. Historically, science and technology have gone their separate ways, but in the recent past the two joined forces in a most effective manner. They are cultivated and bear their fruit in laboratories requiring the cooperation of many people who often come from different disciplines. Research and development on energy will have to be carried out in such big laboratories. This is as it should be, appropriate for the novel and intricate developments that characterize current innovations.

The first change needed is to stop neglect of that research which is so novel that it requires more than three years to bear fruit, and yet so practical that it can make a positive contribution in less than ten years. The question is: How can such a change be carried out? In government laboratories a policy determination suffices. But industry needs inducements.

One inducement would be an improved patent policy. The real purpose of patents is to reward research and, in part, to reward publication about research. The very word "patent" means the opposite of secret. But our present patent law is archaic.

Today a patent is valid for seventeen years; the years are counted from the date on which the patent is granted. I know of one case where a patent concerning electronic computers was based on a claim made before 1950, but granting of the patent was delayed for so long that this rapidly developing field could have been deeply influenced and damaged almost up to 1980. (Eventually this particular patent was held by the courts to be invalid.) Our patent laws were drafted at a time when individual development proceeded at the horse and buggy pace (which was rapid compared to the pace at the time of medieval guilds).

It may be desirable to grant patents for seven years only, and to date them from the time at which the idea was provably conceived. Indeed, priorities of invention are legally based on written and dated proof of the conception of a novel idea.

Of course, seven years is barely long enough to turn a novel idea into industrial practice. But an important and helpful part of present law is that improvements upon patents can in turn be patented. The result is that a well-functioning industrial laboratory can produce patents and improvements fast enough to keep ahead of its competitors. But if industry ever relaxes in research and development efforts, within a few years their patents should become obsolete and their business should be hurt. The proposed changes in patent law would accomplish this.

Such a short-term patent policy should be complemented by liberal behavior on the part of government enterprises. In the past, the Atomic Energy Commission had a policy of denying patents for an invention on which research received any federal support. Formulating a policy that will enable government and industry to cooperate, yet appropriately reward research by the private sector, will be difficult. Flagging industrial research might be accelerated by a sliding formula for government funding of cooperative projects. We may consider the following set of rules as a possible example.

In the initial, relatively inexpensive phases of research, funding would be on a small scale with a large proportion carried by government. To initiate the more expensive second phase, private companies would bid competitively, carrying 50 percent or more of the cost. At this stage an effort should be made to reduce duplicative parallel research. The delicate, difficult part of this proposal would be evaluation of the bids and awarding of contracts to appropriate bidders. The final phases would no longer be research but would be development, in which no direct government grants should be involved.

A direct, straightforward proposal was made by former Vice President Nelson Rockefeller, emphasizing postresearch development. The development should be financed by loans of considerable magnitude that are guaranteed by the government. The amount of capital involved could be $100 billion. The virtue of this proposal is that loans would be granted according to usual financial practices. There would be considerable inducement to select the best applicants because only in that way would a loan bring appropriate interest. The government guarantee would not be employed to insure profits, but would only serve the purpose of reducing losses in case of an unsuccessful enterprise. Since new development is necessarily connected with high risk, the procedure would have obvious advantages.

Another approach may be to remove the barriers against cooperation between industries. Such cooperation today is discouraged by antitrust laws. An important question is whether it is better to have an entire industry cooperate on a new method, for instance a method of finding oil and gas, or to form two or three big cooperative units. It should be remembered that antitrust law originated at a time of great success in American industry, when its rapid development led to excesses that were justifiably criticized. Today the situation is different. OPEC, an international trust, is hurting American interests and American companies. Government–industry partnership in Japan, having all the features of a trust though sponsored by the government, competes with our artifically fragmented industry. However, I do not propose repeal of antitrust laws. I only want to exempt the initial phases, those research and development operations where cooperation may be more productive than competition.

One deterrent to cooperative research and development would be that in cooperative ventures it may be more difficult or even impossible to obtain patents. Actually, patents are useful in stimulating competitive research, but we should not overemphasize the vulnerability that results from the absence of patents. Even after research was fully disclosed by the United States laboratories at Argonne and Oak Ridge, it took years of work by Westinghouse, General Electric, and other companies before the practical production costs of reactors could be reduced enough to make the reactors competitive.

Another difficulty is that know-how is not easily transferred from government laboratory to industry. One possible procedure is to institute training courses. These could be given partly in the national laboratories. On the other hand, it should be possible to transfer groups of experts from government laboratories into private industry so the training and the actual adaptation of a process by private industry occur at the same place.

In all these processes we must recognize the difference between big, established industries and new ventures that are apt to start on a small scale. The former have a considerable advantage in raising capital and in obtaining support within big industrial enterprises themselves. The latter have the advantage of greater flexibility but usually face great difficulties in raising initial capital. Official government certification of expertise of personnel in such small ventures might help in overcoming this difficulty in an important phase of capital formation.

The whole question of research and development is somewhat clouded by a half-recognized change that occurred in the twentieth century. A hundred years ago the concept of an "invention" was generally accepted. That concept survives today even though, on the whole, it has become obsolete. Edison was the last great inventor, and he even invented the end of the inventor. He established the Edison Research Institute, which systematized research on inventions and made of it a highly complex cooperative effort. In the great new fields of nuclear energy, computers, and space exploration there is no "inventor" who builds a new, revolutionary machine such as James Watts' steam engine. Instead, ingenious scientists and engineers with original and unexpected ideas form a team to bring about the miracles of the twentieth century. Our laws, customs, and policies have adjusted to this change in a less than perfect manner.

As with policies regarding pollution and conservation, the policy for research and development cannot be easily or completely formulated. Both policies must evolve through an ongoing period of trial and error. In the case of truly innovative research and development, the trials are more urgent and the errors may be greater. This means only that debate on policies must continue on an empirical and realistic basis. The only point on which I would insist is that the effort must include the participation of both government and private industry. To put excessive emphasis on either would surely result in loss of time, and would neglect the flexibility of our mixed, partially regulated system which has served well in both the distant and the more recent past.

We turn now to a survey of specific alternative energy policies. In the debate on energy policy it has become customary to put one or another solution in the forefront. Some people have advocated that the energy crisis be solved by conservation; others urge the development of renewable energy sources, particularly solar energy; still others favor the development of coal resources or nuclear energy or both. I believe, and have repeatedly indicated in this book, that solving the energy problem will require that all these possibilities be pursued.

I believe we should not place emphasis on any particular type of solution as we formulate a policy. In deciding on a policy we should instead consider several points of view. With respect to the energy shortage it is not intended that we select one point of view as the only relevant one. A compromise among viewpoints may be desirable, but discussing policies from different angles may throw more light on decisions that should be made.

Actually, I propose that we consider four approaches. One I call *business as usual.* Offhand one would say that the very recognition of the energy crisis should exclude business as usual. We should remember, however, that we have practiced business as usual in an acute energy crisis for several years; we should remember also that our economic system has adjusted to many changes in the past.

A second approach I call the *minimum energy option.* This is generally advocated by conservationists. It has an obvious justification in America because America has been traditionally wasteful of energy. In many other parts of the world this approach simply could not be applied. One also may object to it by pointing out a loose but nevertheless real correlation between energy consumption and standard of living (see Chapter 5).

The third course is called the *growth approach.* The rationale in this case is that in the American past rapid development has led to continuing strength, and that loss of confidence in such rapid development may be at the root of present troubles. The obvious and popular objection is that rapid, indiscriminate development will lead to the exhaustion of resources, and will necessarily end in failure.

The last alternative is the *international approach.* This option starts with the recognition that the energy shortage is worldwide, and indeed that it will produce its most disastrous results abroad. Also, the world has become so closely interconnected that no country's development or fate can be considered in isolation. For these reasons this approach is uppermost in my own mind. At the same time I recognize that an international solution will require international cooperation. In the past, constructive international cooperation has been the exception rather than the rule.

We shall make a comparison of the four approaches on the basis of financial requirements for investment and for oil imports. In every case these requirements will be massive. It is remarkable, however, that insofar as they can be foreseen, they will differ very little in each of the four cases. Therefore, we will come to a somewhat surprising result that economic feasibility is not the main argument in deciding which of these approaches should be preferred.

I want to proceed by considering each of these approaches in somewhat greater detail. In doing so, arguments will be given for and against each proposal, and no attempt will be made to make a choice or to rule out any of the options.

First to be considered is the course I have called business as usual. The designation is clear; the implication is usually negative. In times of crisis business as usual cannot be defended.

Yet there are arguments for this option. Five years have passed since the oil embargo. What we have practiced can hardly be called anything but business as usual. So far, it has not led to catastrophic results.

One can go a step further. To consider anything but business as usual may be unrealistic. Our economic system is flexible enough to take care of the disturbances caused by the energy shortage.

Overall consequences of the energy shortage in the past few years resulted in advantages as well as disadvantages to the United States. The latter are obvious: high prices paid for imports and increasing dependence on sometimes unreliable foreign sources. These disadvantages are counterbalanced at least in part by an increasing world dependence on American food production. Thus the balance of payments has not been completely upset; we paid more for oil imports but got more for food exports.

The advantage of the oil shortage to the United States is a competitive one. We are in direct competition with other free, highly industrialized nations such as those of western Europe, some parts of the British Commonwealth, and Japan—countries that also happen to be our allies. As a general rule they are harder hit by the energy shortage than we are. Consequently our economic position is in many respects sounder than it used to be. Our economic leadership in the free world is on a firmer basis. Considering economics alone one may therefore assert that we are in no trouble. The altered conditions should be allowed to produce results in a conventional and relatively slow manner. There is no crisis for the United States; therefore, business as usual is justified.

Another view of the situation is that a restricted energy supply will affect the standard of living in our country. Coupled with continuing demands for greater social justice, this negative effect leads to the conclusion that business as usual is not a satisfactory answer.

Among the options the contrast is sharpest between business as usual and the international option. The strongest objection to business as usual arises from the circumstance that the energy shortage hurts our allies but gives an advantage to those countries with which we may anticipate conflict. An isolationist approach is compatible with business as usual. An international approach is not.

When we evaluate the consequences of this option later in the chapter we shall, of course, rely primarily on current estimates by official sources and business. More or less consciously these sources assume business as usual.

The next alternative to be considered is the minimum energy option. This option is obviously attractive because it solves two problems and tends to solve them in a durable fashion: one is the energy problem itself; the other is the problem of environmental pollution. If we burn less fuel, pollution will be reduced.

Furthermore, the minimum energy approach is the plausible one because of the obvious fact that energy is used wastefully in the United States. As mentioned earlier, in Switzerland and Sweden, which have standards of living similar to that in the United States, per capita energy consumption is considerably less than in the United States.

These arguments for using less fuel are strong. They are also incomplete. It has been pointed out that significant energy saving can be achieved in only two ways: by a radical change in people's habits, or by more investment in energy-saving devices. Complete and effective enforcement of the former would require centralized dictatorial powers. The latter would have serious effects upon our economy.

As pointed out in Chapter 5, a direct comparison between Switzerland, Sweden, and the United States is misleading. It neglects our size and our unique role in the world. If the free advanced democracies were all like Switzerland and Sweden, the free advanced world might no longer exist.

Energy conservation is an obvious necessity. The strongest objection to the minimum energy option is that it assumes that conservation alone, without efforts to increase our energy supply, can solve the problem. Adoption of this approach would mean that measures should be taken not only to save energy but to discourage introduction of new energy sources. This seems arbitrary and unnecessary.

The minimum energy option does not mean that research on novel ways to generate energy should be ruled out. On the contrary, such research may lead to the possibility of reasonable restraints on wasteful energy plants. They should be replaced by more economic ones. If one wants to pursue this option it is necessary to do so in an imaginative manner.

The third approach on our list of four, the growth option, assumes that we return to traditions of the past. This does not mean that the growth option is conservative. On the contrary, this course requires rapid change not only in producing more energy but also in developing novel energy resources so that growth can be sustained over a long period of time.

This option runs contrary to a widely accepted sentiment: that continued growth will lead to trouble and is in fact evil. There can be little doubt that very many people in the United States believe strongly in the truth of that statement. There are also many who are convinced that it is completely false.

It is clear, and it must be reemphasized, that the growth option cannot be ruled out simply on the basis that it would quickly deplete our resources. New resources can be developed and made available.

It is equally clear that the growth option would require a change in our attitudes and probably the reduction or even the abolishment of present governmental regulations. In fact, in the United States governmental regulations have proliferated in response to demands to eliminate malpractices and to further social justice. In recent decades our government has not espoused the cause of national growth. In this regard the Soviet Union is our complete opposite; indeed, in the Soviet Union governmental efforts have been directed primarily toward growth.

As with the other options, the growth option can be adopted and can succeed if there is a national determination to make it succeed. As with the other options, a price must be paid for success.

The growth option would necessitate our pursuing all promising energy developments to the maximum feasible extent. This would include new approaches to energy production. It would not, however, emphasize energy conservation. In the American experience it has always proved more profitable to find and use new energy sources than to economize. Even though this tradition has been broken, no discussion would be complete without consideration of the effects of reestablishing it. Accordingly, it would be important to estimate what reasonable limits should be set to energy development. These limits would then depend on what is effective in terms of human effort; they should not depend on preconceived notions that growth must be limited. All this should not be construed as a plan to which we must adhere, but only as an estimate of what might be accomplished. Adherence to a plan would contradict the spirit of the growth option.

Finally, we come to the international option. The energy crisis is international. It cannot be solved except on an international basis. The world has become small and interdependent. Even if the United States solves its own energy problem we shall suffer from the damage that the energy shortage will cause abroad.

The main argument against the international option is that it is unrealistic. It is hard enough to solve problems in one country. Some believe that our country is so big and so complex that even within the confines of the United States problems become insoluble.[7] From this point of view it will seem hopeless to tackle the energy problem on an international scale. Indeed, international agreements are singularly hard to achieve and harder to enforce.

Nevertheless, I see some hope. It is based on the fact that America is rich in natural resources and has a long and strong tradition of industrial innovation and development. The very existence of the worldwide energy shortage opens the possibility for increased American influence throughout the world by the peaceful means of providing energy resources where they are most urgently needed under conditions benefiting both the recipient country and ourselves. A similar situation has already arisen in agriculture.

If the United States could generate energy surpluses as well as food

[7] I am paraphrasing here a statement made to me by Antony Wedgewood Benn, Secretary for Energy in Great Britain. He did not happen to mention that the Soviet Union may suffer from the same reckoning and the same difficulty.

surpluses, and if those surpluses were used wisely, the chances for constructive international cooperation would be greatly enhanced.

The trouble with this option is that Americans themselves will probably not adopt it. It presupposes that we are willing to act in some cases against our short-range interests, and to be guided by results that lie farther in the future. To economists it appears too costly; to environmentalists it seems too dangerous and dirty.

The proposal would require full exploitation of American oil resources, which may near exhaustion by the end of this century or the beginning of the next. We can pursue such a policy only if we have confidence that other energy resources can be developed soon.

Caution seems to argue against the international approach. But an energy shortage is not the only challenge we face. Those who want to put American interests first, overriding the interests of others, may believe that they are cautious. From a broader point of view they may be playing the most hazardous of games.

From the aspect of internal requirements the international option is similar to the growth option. But there is a difference. In the latter no plan is involved; in the former it is preferable that we know where we are heading. We are considering here growth with a purpose. One explicit difference is that in the growth option there is no special emphasis on conservation; in the international option there is. The more we conserve, the more we can help others. The more we conserve, the more we can teach others how to conserve.

Having defined the options one by one, we turn now to a comparison of the four alternatives. A clear analogy exists between tbe present situation and World War II. In the intervening three decades no danger has been as great as the one now developing. Perhaps the analogy with the Manhattan Project, which has been mentioned many times, is not quite appropriate. World War II was won by the Manhattan Project, by radar, by liberty ships, by producing fifty thousand and more fighter planes per year, and by the bravery of many individuals. The problems that we now face will most certainly not be solved in any one specific "Manhattan Project" way. I hope that in the troubled times that are undoubtedly coming we shall never reach violent action or the need to dislocate and regiment many people. If we choose the right alternative or the right combination of alternatives we can avoid real trouble.

It is proper to describe alternatives by estimating how much they cost. I shall discuss the needed energy-related expenditures for the next ten years. These expenditures include capital for new facilities.[8] From this we should subtract return on the investment during the ten-year period. We should also add the cost of oil imports. Since these oil imports are not completely dependable it is advisable to store oil for six months or more. Energy-related expenditures for environmental protection must be included. And finally we include the research and development expenditures connected with energy. All these particulars are summarized in Table 13-1.

There are objections to this kind of "bookkeeping." For instance, it is clearly erroneous to consider the cost of construction in the United States under the same heading as the cost of imported oil. The former is under American control; the latter is not, though the money might be reinvested in the United States. We have subtracted an appropriate return on the investments expected within the United States. Even this does not fully convey the advantage of internal investment, since such investment creates jobs and strengthens the economy.

The major point is that total net energy-related expenditures for the next ten years will be between $500 billion and $1 trillion, which is approximately one-third of the annual Gross National Product. If we count expenditures alone, without return on capital, the figure is approximately $1 trillion.

Another, more striking comparison is that these net expenditures will exceed 10 percent of our total national worth, including such items as the value of minerals known to be underground and which have been left there. The magnitude of the financial requirements is so great that it will deeply influence and might conceivably break the fabric of our economy. Under these conditions it may seem proper to lump together all these financial requirements and to compare them for the options that we have discussed. We also compare them with the figures of net energy-related expenditures for the past decade, 1966–1976, which total, as we see, is little more than $100 billion, or about one-sixth of the figures projected.

Of course, all of the projections are guesses. What is worse, they are guesses that cannot be confirmed, because in reality some compro-

---

[8]This estimate should be corrected by adding money spent on facilities that are started but not completed in the ten-year period. One also should subtract expenditures made up to the present on facilities that were started in the previous ten years but are not yet finished. The former item is apt to be bigger, but the difference is within the uncertainty of the estimate we are making, and is neglected.

TABLE 13-1 *Energy-Related Expenditures, in Billions of 1977 Dollars*

	1966–1976	*1977–1987 Business as usual*	*1977–1987 Minimum energy*	*1977–1987 Growth*	*1977–1987 International*
Electric utilities	84	200	200	320	320
Petroleum	100	160	160	300	300
Coal	5	20	20	40	40
Conservation	10	40	160	40	160
Return on investment	−81	−170	−260	−280	−370
Environment	10	50	50	50	80
Research	20	20	50	100	150
Imports	70	500	250	250	120
Storage	0	10	10	20	5
Gross expenditures	199	1,000	900	1,120	1,175
Net expenditures	118	830	640	740	705

mise between the options will be found. Obviously it is necessary to explain the basis of these figures, the more necessary because they are uncertain.

The two biggest investments that will dominate the expenditure picture are electric utilities and the capital needed for petroleum. In the past ten years we have spent $84 billion on the former and $100 billion on the latter. It is obvious that in the next ten years these figures will be greatly increased.

The cost of electric generating plants has more than doubled in ten years. Most estimated figures from official sources and from the private sector lie between $200 billion and $300 billion. I use the $200 billion figure for the business as usual and minimum energy options, but $320 billion for the growth option and the international option. The latter figure happens to agree with the maximum used by the Federal Energy Administration in 1975. (An adjustment from 1975 to 1977 dollars would actually increase the Federal Energy Administration figure to $340 billion.)

Estimates for petroleum expenditures are generally similar. For the business as usual and minimum energy options I adopt $160 billion, the estimate of the Federal Energy Administration in case ceiling prices on oil and gas are maintained. For the growth and international options I adopt $300 billion, which is close to the maximum estimate.

It will be noted that in the ten-year period capital for electricity is assumed to increase less than capital for oil and gas. Recent trends make it necessary to decrease the use of oil and gas, which will not continue in ample supply. Electrical generating plants will actually shift from oil and gas to coal and nuclear energy. Differences between the increased capital needs for electricity and for petroleum would be even more marked if it were not for the fact that changes cannot occur in a few years. The lead time for new plants, as well as for new oil and gas production, is in the range of five to ten years.

I have used the same capital investment figures for business as usual and minimum energy. In the case of business as usual I am extrapolating from the present situation in which capital investment is low. The main difference between the minimum energy option and business as usual is that the former stresses energy conservation.

The situation is analogous for the growth and the international options. In both cases development in the United States of electric generating plants and investment in petroleum are to be driven ahead as fast as it seems feasible. The reasons for this development in these two options are different, however, leading to different consequences in other energy investments.

In the past ten years we have invested only $5 billion for coal. Indeed, we have tended to neglect coal mining in this period, partly because oil has been cheap and partly because regulations affecting the coal industry have been tightened. It is obvious that coal production will be greatly increased. We use the same figures, $20 billion for business as usual and for the minimum energy options; this figure is four times the amount expended in the last decade. In the growth and international options we use $40 billion, eight times the figure for the 1966–1976 period. Capital investment for coal is nonetheless low. Production need not be limited by the availability of capital, but rather by the amount of coal that can be used.

It is estimated that in the past ten years $10 billion were spent for energy conservation, most of it actually in the last two years. This figure is uncertain and hard to estimate because it has been spent throughout industry as well as in individual actions such as bettering home insulation. For the next ten years I adopt the figure of $40 billion for business as usual and for the growth options. In both cases the rate of investment of the past two or three years is maintained. In the case of the growth option I use this figure because attention will focus on production rather than on conservation. On the other hand, for the minimum energy option an investment of $160 billion is used for the coming decade. For this option conservation is indeed the main point. The same figure is used for the international option because in this case we do not merely want to satisfy internal demand, we want to have as great an influence as possible on the rest of the world. The purpose of conservation would be to decrease imports, and turn them into a surplus of energy exports toward the end of the period. In this way we could have a decisive influence on the world market and could make a considerable contribution toward solving the worldwide energy crisis. In the present discussion we consider only conservation accomplished by the introduction of energy-saving devices. There would be important added conservation if people actually use less energy. In fact, the two categories of saving cannot be clearly separated because attention to conservation will affect decisions on individual use of energy as well as introducing appropriate equipment.

From these items we should subtract the return on the various investments. Consequently, the figures appear with the negative sign. These figures were obtained by assuming that half of the investment in oil and gas and in coal will be repaid within the decade. Electric utilities are regulated to limit the return, and at present they are in a poor financial condition; it is therefore assumed that only one-fourth of the investment is returned. Conservation, on the other hand, seems to be highly profitable (that it is not practiced to a greater extent is due to deeply rooted habits). I assume that in ten years three-fourths of the investment will be repaid. The figures so calculated are rounded off and the sum is entered as return on investments.

It is hard to estimate expenditures on energy conservation. It is much harder to find proper figures for the expense of environmental improvements. In fact, if the proposal to "internalize" environmental costs were completely adopted, the expense of environmental improvement might not even be explicitly shown and might become impossible to trace. There would seem to be an advantage in carrying environmental improvements as an explicit and separate expense. It is obviously necessary to have legislation to enforce an appropriate environmental standard. But it also seems reasonable that we should recognize how much enforcement of this standard increases costs. Even more than with other expenditures the cost of environmental safeguards is uncertain. It is hardly more than guesswork.

In this sense I am guessing that in the previous ten years we spent $10 billion on energy-related environmental measures. This amount includes direct and indirect costs of antipollution measures for automobiles, added costs carried by utilities due to restrictions on emission from stacks (restrictions on fuels or clean-up procedures of the stack gases), expenditures connected with location of electric generating plants and refineries, and furthermore, expenditures to repair damage from strip-mining. The actual expense may have been larger. In many cases plants or coal mines have just not been developed. Therefore the economic impact of environmental restrictions is much greater than represented by the $10 billion.

There is little question that environmental problems will have to be resolved, and are in fact being resolved. The result will be increased expenditure, hopefully coupled with some increase in freedom of action. Indeed, development of clean energy resources is feasible at an appropriate price. Unless extremists prevail, the bottleneck in siting power plants, if not broken, will at least be less narrow and constipated. My guess of the environmental cost for the next ten years is $50 billion for business as usual, since environmental legislation is an accomplished fact and business will have to adjust, and the same amount for the minimum energy option, where actual deployment of energy plants is similar to business as usual. The same amount is estimated for the growth option, where more energy-producing plants are deployed, but in the growth option spirit, less emphasis is placed on regulation. For the international option I am using a figure of $80 billion, which is the figure employed for business as usual and minimum energy multiplied by 1.6, the rough ratio of investments for energy-producing plants.

The energy shortage demands the introduction of new energy sources. Research and expensive development have proceeded, and for the previous ten-year period we may estimate its cost at $20 billion. The bulk of industrial research is not included in this figure because previous estimates hardly ever included plans beyond three years. Such short-term expenditures are better considered as operational expenses rather than developments for the future. Long-range developments have been mostly carried out by the government. In the business as usual option it would seem reasonable to retain this $20 billion figure for the next ten years. For the minimum energy option new methods are needed, particularly for developing energy-conserving equipment. For this reason a figure of $50 billion is used. In the growth option the figure is increased to $100 billion, and in the international option to $150 billion. In these latter two approaches there is increasing need to meet greater energy demands and, in the case of the international option, to meet them in such a manner that growth can be applied throughout the world.

The immediate cause of the energy shortage is, of course, the oil situation, most clearly and painfully expressed in the circumstance that in 1977 the United States spent over $40 billion on oil imports. Seventy billion dollars was spent in the preceding ten-year period. It is obvious that in the business as usual option expenditures for imports will not only continue but will increase. A relatively moderate estimate is that if we continue present practices, $500 billion will be spent for oil imports. This one figure by itself justifies the variety of radical measures we are discussing. In the minimum energy option we may hope to reduce this figure to $250 billion by energy-saving devices and individual energy-saving practices. In the growth option I assume that imports will cost a similar $250 billion, the reduction due to increased domestic production. In the international option the net imports could be held to $120 billion. This figure would result from continuing but diminishing imports for a number of years, followed by a shorter period in which we earn more by energy exports (coal and nuclear) than we pay for oil imports. In fact, at the end of the ten-year period our oil imports might be eliminated.

According to international agreement and domestic legislation we are to establish oil storage in order to decrease the impact of a renewed oil embargo upon us and our allies. A small part of the expenditure will pay for storage space, a greater part for the stored oil. In the previous ten years a negligible amount was spent on storage. In the business as usual case I assume an expenditure of $10 billion, which should permit oil storage. This is indeed a minimal amount to be spent for compliance with the treaty and the law now in existence should we continue or increase oil imports. A similar expenditure is used for the minimum energy option, but here the same fuel will suffice for a longer period, since less imports are foreseen. The growth option aims at a strong United States position, so an expenditure of $20 billion is estimated, allowing us to forego imports for more than a year.

Finally, in the international option $5 billion is assumed for oil storage. We should realize that storage facilities will not become instantly available. If we adopt the international approach and aim at elimination of oil imports, they will already be decreased substantially by the time fuel storage can be fully deployed. Therefore, spending as much on storage is not justified in this case as in the growth option.

The last two lines of Table 13-1 give the total gross expenditures, that is, the sum of all figures without deducting return on investment, and total net expenditures, that is, the total after return on investment is subtracted from the gross. It should be noted that while expected net expenditures are all below $1 trillion, the gross expenditures are actually in the neighborhood of a trillion dollars. Some consider the gross figures more appropriate, since they do not involve assumptions about return on investments.

With expenditures spelled out, the differences between the various options become clearer. One additional consideration is significant. In the case of business as usual we assume that oil and gas prices remain regulated in roughly the same way as they are now. In the minimum energy option price increases will reduce consumption, which is that option's purpose. On the other hand, the increased price would stimulate production, which is the goal in the growth option; in this case the need for deregulating prices is obvious. One may note an essential difference: in the minimum energy option more capital is channeled into conservation, while in the growth option more capital is employed in expanding production capacity. In the international option even more capital is spent on both conservation and production capacity. The hoped for result is that this will reduce imports and have a beneficial impact beyond the borders of the United States.

In estimating that most important cost, net imports, it has been assumed that the price of oil will remain at the level of $13 a barrel in 1977 dollars. There is, of course, no assurance that this assumption will hold, and in this case we are facing a situation more complicated than mere uncertainty. Some have argued that oil prices will drop. This would make business as usual more attractive because in that option most money was assumed for expenditure on imported oil, and most could be saved if the price goes down. On the other hand, business as usual is the approach requiring the greatest oil imports, which is apt to drive up the oil price. If we adopt that option the imports may turn out to be even more expensive than projected in the table. However, if we follow the international option the oil price might come down. That would make little difference for that particular option; but if all this comes to pass there will be those who argue that we should have adopted the business as usual position, by which we would have profited most by decreasing prices. This means no more than the usual supply and demand dilemma complicated by the circumstance of long lead times.

The foregoing holds if the United States alone is considered. But our behavior carries worldwide impact due to the size and potential strength of our economy. The international option creates meaningful competition with the Organization of Petroleum Exporting Countries. It seems to be the only way in which, in the long run, the imbalance created by OPEC can be adjusted.

Two details on the table of expenditures are remarkable. One is that expenditures run very much higher than they did in the previous ten years. This is not a consequence of any decision that can be made now; it follows from mistakes committed in the not-so-distant past.

The other point is that the high expenditures we have to face are similar for the four options. If we consider total net expenditures, they are highest for business as usual, next highest for the growth option, lower in the international case, and lowest for the minimum energy approach. But the highest figure is only about four-thirds of the lowest. The uncertainties of the prediction are greater than the differences.

None of these conclusions is too surprising. Our present economy contains elements of self-adjustment so that no change in action will result in dramatic financial consequences. This is as it should be in a balanced economy. If a change could have profited us greatly, it would already have occurred.[9] The one exception is the minimum energy option, where we could profit by breaking our wasteful habits. But we should not overestimate the difficulties in breaking any habit.

We may conclude that the decision of which option to choose or how to balance the procedures is less a question of economic necessity than a matter of choice and taste. OPEC's success depended on natural inertia, the consequence of business as usual. A radical departure from this pattern, following the international option, could result in a permanent solution of the energy shortage. A return to older American traditions in the growth option could meet our domestic needs, and the minimum energy option would require an heroic effort to break age-old national habits.

The advantage of figures is that they are clear and definite. The disadvantage of figures is that they are too clear and too definite. We ignore them at our peril and we rely on them at our peril. We may discuss them with profit and without end. They can serve as a basis for a plan of action. At best this plan will be flexible and should be changed according to necessity.

---

[9]Unfortunately, the last statements may be incorrect. They may even fail to reflect the real dangers of some radical change, for instance, a collapse of the economy. This is a possibility that may occur in any of the alternatives discussed, and that is likely to be influenced by factors other than energy supply. Prejudices of the reader will have a lot to do with his judgment about which of the alternative catastrophic events is most likely.

CHAPTER 14

A MODEL FOR THE FUTURE

In which the reader will find that adolescence, the present stage of energy's development, is a time when advice and predictions are amply available, though not always requested or helpful. An attempt is made to see how promptly troubles can be resolved, and what might happen in the remainder of the century in the United States and in the rest of the world.

What is a model?

Webster's *New International Dictionary* has fifteen definitions of a model, both current and obsolete. What I mean here does not really agree with any of Webster's definitions.

A few decades ago when ideas about atomic structure were still full of contradictions it was popular and useful to talk about models of the atom. The model consisted of a set of ideas that gave rise to clear-cut mathematical relationships. Although the model was expected to behave like an atom in important ways, one did not really believe that the atom looked like the model.

Economists, on whose difficult field I am encroaching with a feeling of some trepidation, started in the sixties to use a similar concept of a model: a number of ideas simple enough to permit comprehension, expressed if possible in a mathematical manner. To the extent the results agree with reality the model is successful. Complete success would amount to a miracle.

The sense in which I am using the word model here is similar to that in which physicists or economists use it, in that the model is not expected to describe reality. Further, it is similar in that the model is constructed with the hope that it will be useful to compare it with reality that develops in the next decades. However, the model I am now discussing is not clear nor logical enough to lead to any mathematical results.

I certainly cannot predict the future. I also want to avoid talking about a plan for the future. A plan would carry the implication that what is suggested should be enforced in some way and thereby turned into reality. By discussing a model, however, we can visualize in what way the future might develop.

In Chapter 13 we discussed alternatives for constructing a model. I shall ignore the alternative called business as usual, partly because I consider it undesirable to continue business as usual in the field of energy, but also for a second and even stronger reason. We know what business as usual is because we are experiencing it. The model that I shall discuss differs from past experience and therefore should be more instructive, though not necessarily more accurate. I am less interested in being right than in stimulating ideas on the subject.

Among the other three alternatives—the minimum energy option, the growth option, and the international option—I do not wish to make a choice. Instead I shall propose a model that incorporates features from all three. It is certain that the actual future will not be what is here described, but the future might be influenced in some small measure by how people think about it.

We shall start by comparing actual data on energy demand and supply with possible future situations.[1] If there is one thing I am certain about it is that I am not a prophet. The future depends on the unpredictable ways in which our own society develops and the manner of our interaction with the world. It depends on the vigor and success with which we pursue technical developments. It depends on the question of whether that remarkable cartel called OPEC will continue to function, in defiance of the American antitrust laws which do not apply, and also in defiance of an imagined law of economics according to which producers' cartels cannot work. But with all these uncertainties some shifts in the energy picture can be foreseen, and the figures given below have the purpose of giving an impression of the nature of these shifts.

The last year for which reasonably complete data are available to me is 1976. The model will compare future energy demands and energy supplies with these data.

It is clear that the energy problem demands urgent action. In fact, within five years, by 1983, the situation could well turn desperate. What might reasonably be accomplished by that year will be compared with the 1976 situation. A short-term discussion has the further advantage that the uncertainties are not too great.

For the more distant future, the year 2000 is being considered. By that time many new plants including electric generating plants will be built. Policies which at the time of writing are still being discussed, concerning oil, gas, and coal, will have had full opportunity to make themselves felt. And by that time results from newly developed technologies can begin to play a role.

---

[1]During the years 1973–1975 I was a member of the Committee on Critical Choices for Americans, founded and chaired by Nelson A. Rockefeller. In that period I worked out and published a proposal, "Energy—A Plan for Action," similar in subject matter to this chapter. The results of that paper were quite different from those presented here. For the year 2000 I projected an energy output almost double that calculated in this chapter. One reason was undiluted optimism based on ideas of what could happen in the best of all possible worlds.

My optimism is now diminished due to three facts. The first five years after the oil embargo have been effectively wasted. Prospects for finding more oil and gas in the United States have become less promising. And a powerful antinuclear movement has arisen in the United States as well as abroad. However mistaken the reasons for this movement may be, its effects are undeniable.

Table 14-1 below, concerning energy in the United States for the years 1983 and 2000, has some of the beauties of a balance sheet that can be read only by those who expect what they will see. Therefore, it is necessary to transform (painlessly) the reader into an expert.

The figures in the table are given in a peculiar unit called a quad.[2] A quad is a quadrillion British thermal units (1,000,000,000,000,000 Btu, or 10^{15} Btu).[3] The reader may get an idea for comparison of that unit from the statement that the United States used approximately seventy-five quads per year in the last few years, while the whole world used about three hundred. An average American household uses approximately one-millionth of a quad per year. Indonesia, however, with a population in excess of a hundred million, uses between two and three quads per year.

In regard to crude oil[4] we have become accustomed to speak of millions of barrels per day, due to the repeated shocking news of how many millions we imported in each of the past few years. It is therefore useful to know that a million barrels per day is the equivalent of 2.12 quads per year, and conversely a quad is equivalent to .472 million barrels per day. Natural gas, or equivalent synthetic material, is usually measured in commercial statements in trillions of standard cubic feet. It simplifies the situation that the energy delivered by a trillion standard cubic feet is just one quad.

The biggest and most peculiar single entry in the table is the first one: Electricity for one year. Under this column we will not list the energy delivered by electricity, but rather the energy used up in generating electricity. The electric energy generated is less than the energy used by the ratio called efficiency, which at present is approximately one-third.[5]

Energy consumed by people in their homes and places of business appears under the heading Residential/Commercial. The contributing en-

---

[2] In a discussion on energy a famous, charming, highly intelligent banker asked me, "Aren't quads what Brazilian women get when they are pregnant?"

[3] If energy could be converted into matter (impossible according to present theory unless we get hold of some antimatter) one quad would be generated by the annihilation of approximately twenty-five pounds of matter.

[4] Crude oil is, of course, the source of gasoline, kerosene, diesel fuel, and oil residues.

[5] So far the procedure may be considered strange but at least logical. In fact there is a flaw in the logic. Under hydroelectric power we enter the fuel that would have been used up had the electricity been obtained from coal or oil. The actual energy contained in the fall of water is little more than the energy developed as electricity. The inconsistency is tolerated because the hydroelectric contribution is relatively small.

TABLE 14-1 *Actual and Projected U.S. Energy Demand and Supply, in Quads*

Demand	*1976*	*1983*	*2000*
Electricity			
Total	*21.4*	*25*	*45*
Oil	3.5	3	0
Gas	3.1	2	0
Coal	9.7	12	10
Nuclear	2	4	20
Hydroelectric	3	3	5
Solar	0	0	1
Wind and waves	0	0	3
Burning of Wastes	0	0	2
Geothermal	0.1	1	4
Residential/Commercial			
Total	*13.6*	*10*	*10*
Oil	5.3	4	3
Gas	8.1	6	4
Coal	0.2	0	0
Solar	0	0	3
Electric	[4.1]	[5.5]	[10]
Industry			
Total	*15.4*	*18*	*25*
Oil	3.9	3	4
Gas	7.8	7	9
Coal	3.8	8	10
Solar	0	0	2
Electric	[2.8]	[3]	[10]
Transporation			
Total	*19.2*	*18*	*16*
Oil	18.6	17	12
Gas	0.6	0.5	0.5
Other Fossil	0	0.3	1.5
Biomass	0	0.2	2
Electric	[0]	[0]	[3]

TABLE 14-1 *Continued*

	1976	*1983*	*2000*
Nonenergy Demand			
Total	*4.5*	*5*	*10*
Oil	3.7	3	0
Gas	0.7	1	0
Coal	0.1	1	9
Biomass	0	0	1
Total Domestic Energy Demand	74.1	76	106
Oil	35	30.3	20.5*
Gas	20.3	16.5	13.5
Coal	13.7	21	29
Other	5.1	8.2	43
Total Domestic Energy Supply	59.8	69.2	112
Oil	20	23	22*
Gas	19.4	15	12
Coal	15.3	23	35
Other	5.1	8.2	43
Total Imports	14.3	6.8	−6
Oil imports	15	7.3	−1.5
Gas imports	0.9	1.5	1.5
Coal exports	1.6	2	6

*Amount includes liquids derived from other fossil sources (oil shale, coal liquefaction).

ergy sources add up to the sum that appears on the line with the word total. Electricity is not counted among these items, however, but appears in the table in brackets. The figure in brackets is the electric energy delivered. Therefore, if we add up all the bracketed figures in one column, representing electric energy used in one year, the sum will be less than the total figure given for energy needed to generate electricity. The ratio, of course, is the efficiency.

Industry is carried in a separate heading because, in this case, the main use of energy is not to satisfy people's momentary needs but rather the production of goods. Electric energy is again shown in brackets.

Under the heading of Transportation we have set the electricity consumed equal to zero for the years 1976 and 1983, though electricity plays a minor role in mass transportation called rapid transit systems.

An item called nonenergy may appear puzzling in an energy model. What is meant is the use of raw materials such as oil, gas, and coal, for purposes other than energy production; this category includes anything from asphalt used in road-building to fertilizers, toothbrushes, and paint.

The last three headings are Total Demand, Total Domestic Supply, and Imports or Exports, whichever predominate as a rule. The latter represent the difference between demand and supply as far as coal, gas, and oil are concerned. A negative sign in an imports column indicates that exports exceed imports for that year.

Figures have two important properties; they say a lot and they appear to be accurate. In both respects there is a real disadvantage connected with the apparent advantage. Figures say so much that without a detailed discussion it is impossible to understand them. When the discussion is furnished, the substance is in the words rather than the figures and the latter are only a device by which to remember the discussion.

Figures are also much too accurate. The trouble is that with regard to the future this accuracy can never be attained.

These considerations must be kept in mind as we now turn to a detailed discussion of the figures in the table.

The reader will probably first look at the bottom of the table and find that we imported more than 14 quads of energy in 1976, which corresponded to seven million barrels per day and cost over $32 billion. According to the table, oil imports should fall to less than one-half by 1983. This will require an heroic but necessary effort.

One important portion of this effort is to save energy. Indeed, according to the table, we hardly permit the per capita energy consumption to rise.[6] Another part of the effort is a vigorous increase of coal production, by 50 percent,[7] and the pursuit of our plan to build nuclear reactors whose output by 1983 should be doubled.

---

[6] A modest increase of total energy production is required if our industrial strength and our ability to play a positive role in the economic development of the world is to be maintained.

[7] This assumes no crippling strikes. It should be noted that working conditions in eastern underground coal mines are hard and strikes are more easily understood. Development of surface mining in the West should alleviate the situation.

As a consequence of these changes, domestic demand for oil should decrease from 35 to 30 quads, and the per capita consumption of oil should be reduced by 20 percent.

But not even all these efforts suffice. It is also necessary to increase the domestic supply of oil from 20 quads to 23, a 15 percent rise. This increase is possible only if exploration for oil is vigorously and systematically encouarged. It is in this respect that the greatest change in current public attitudes is necessary. To deregulate oil is one of the required steps, but it is not the only step we will have to take. Without encouraging domestic suppliers of oil, we shall of necessity encourage the OPEC cartel and contribute to the worldwide oil crisis.

Apart from the difficulty of effecting the necessary changes in public perception, many will believe that the proposed early increase in oil production will, in the long run, lead to the exhaustion of our oil supplies. Present estimates show that the proposed oil production for the year 1983 could be maintained until the turn of the century, and would decline thereafter, steadily but not rapidly.

In this connection it is of interest to look at our table for the year 2000. Here the table shows that our demand for oil will have declined from 30 quads in 1983 to approximately 20 quads by the turn of the century. The decline is accomplished by the following detailed measures.

☐ In the seventeen-year period the growth of energy consumption is limited to 2 percent (the per capita consumption will increase a little less rapidly). Such an increase, approximately one-half of the historic figures preceding the seventies, is compatible with a strong industrial and international position only if continued efforts are made to save energy.

☐ Coal production is assumed to increase further in the seventeen-year period by more than 50 percent. The final result would be more than twice our present coal production. The need for such a change is recognized, and there are no obvious and immutable political objections.

In the table it is assumed that our production will rise even more steeply, so as to give a chance for the export of coal. This can be done without excessive capital investment and without the danger of an early exhaustion of our coal fields. Coal export could be an important contribution in alleviating energy shortages abroad.

☐ Electric generation by nuclear reactors is supposed to increase more than fivefold in the years between 1983 and 2000. At the latter date, nuclear reactors alone would produce as much electricity as we use today. They would deliver almost half of the electricity consumed in the year 2000. This cannot happen unless the public thoroughly understands the fact that nuclear energy is clean, safe, and relatively inexpensive.

☐ All forms of solar energy will contribute almost as much as nuclear reactors. Approximately 30 percent of this solar energy is familiar hydroelectric power. The rest, well in excess of 10 quads, is obtained from novel technologies. This will require effort and luck.

If we maintain production to the extent indicated in the table, in the year 2000 we could export 1.5 quads equivalent of oil. At later times, due to further application of new technologies including nuclear reactors, our demand for oil could continue to fall. Even as a fuel for automobiles there is a good prospect for replacement of gasoline by products obtained from biomass and oil shale.

One way to view the big changes in the energy picture is to see what percentage of energy needs will be covered by the "classical" sources: oil, gas, and coal. For 1976 it was 93 percent, for 1983 it would be 87 percent, and for the year 2000 nearly 60 percent.

We must recognize that most electric power stations are being opposed at present by some extremists in the environmental movement. It is of some interest to see how their arguments would apply in the year 2000.

Of the 45 quads devoted to electricity, the 10 quads for coal may well be the most polluting. The amount compares with 8.8 quads of coal for production of electricity at present. With vigorous research on cleaning up the emitted pollutants and the use of coal with lower sulfur content, of which plenty is available, the pollution caused by coal could be substantially decreased.

Furthermore, we should consider that cogeneration will play a much more important role by the year 2000. In this sense it is somewhat arbitrary to say that 10 quads of coal are used for electric generation and 10 quads, as shown in the table, for industrial use. It might be more accurate to say that 5 quads of coal will be burned in generating stations of utilities and 30 quads of mixed fuels will be burned by industry (4 quads from oil, 9 from gas, 2 from biomass or solar origin, and 15 quads from coal). Of these 30 quads, industry may use 25 for its own purposes while 5 quads find their way back into the electric distribution system because of the wide introduction of cogeneration. It is in this somewhat loose sense that we assign 25 quads of heat to industry and 5 + 5 quads to generation of electricity from coal, with the second 5 quads derived from cogeneration, not all strictly due to coal.

The burning of wastes could be as polluting as coal, or more so. We must remember, however, that most of the wastes constitute pollution. If the process is brought under proper environmental control, the harmful emissions might be reduced in the same way as those from coal.

The next most polluting source probably is the 4 quads assigned to geothermal energy production. Gases are discharged with the hot water, such as hydrogen sulfide, which could become a severe pollutant. Preventing this, and disposing of wastes carrying salt and silt, are among the problems that have to be solved if widespread use is to be made of geothermal electricity. The 5 quads from hydroelectric power and the 20 quads from nuclear power are nonpolluting.

Three quads from wind and one quad from solar electricity will be nonpolluting and essentially not dangerous. Some complaints have been heard about visual pollution.

It is unclear whether it will be possible to increase the energy devoted to electricity from the present value of 20 quads to a prospective value of 45 quads. Such an increase may mean serious difficulty due to the great capital expenditure required by electric generation. Competent studies have indicated that gas obtained from coal gasification will be a less expensive form of energy than electricity. The comparison may further shift in favor of gas if present expensive methods of gas production could be replaced by cheaper methods of underground coal gasification.

The trade situation in oil is of course the most important; gas and coal will become important. Gas imports from abroad, in the form of liquified natural gas, are increasing. Liquefaction and transportation, however, are expensive, and the material itself is dangerous. Hopefully, this form of gas importation will not become a larger item in our trade balances. However, importation by pipeline from Canada and the new fields of Mexico may be reasonable. I have assumed that domestic demand which, according to the table, would decrease by the year 2000 to seven-tenths of the present demand, would be satisfied primarily by domestic supply, but the importation of 1.5 quads of gas is assumed. The uncertainty of such a prediction is obvious.

In comparing the cost of gas with that of electric power we should take into account that its developing forms could also become less expensive. Radical reduction of cost for nuclear power could evolve from standardization, simplified licensing procedures, and possible introduction of methods of mass production whereby reactors are produced in central locations and then deployed on barges.

It is also important to improve the efficiency of generating electricity. In 1983 the electrical energy delivered (5.5 + 3 = 8.5 quads) will slightly exceed 33 percent of the original 25 quads used in the generating process (we continue to disregard the inconsistency connected with hydroelectric power). In the year 2000, according to our model, the efficiency will approach 50 percent. By that time, electricity delivered in the residential/commercial and industrial sectors will total 10 quads each. A further 3 quads appear under Transportation. Indeed, some of the transportation will be electrified, not only in mass transportation, which has been neglected in the 1976 and 1983 entries, but also by the possible use of electric cars or cars with auxiliary electric power.[8]

Comparison of gas with electricity is only one of the examples showing how unpredictable the future of energy is. To place more emphasis on one or the other developing technology could radically change the appearance of our table. However, the general conclusion that we can become self-sufficient in energy need not change. The table's construction was guided by emphasizing known technologies and giving a second, but still important, place to developing and popular new proposals.

The table shows electricity to have a peculiar role. At the present time the generation of electricity corresponds to less than 30 percent of our total energy demand; by the year 2000 the electricity demand would be a little more than 40 percent of total demand. These changes correspond to an historic trend. Use of electricity has increased faster than use of other forms of energy. The growth rate assumed here is, on the average, about one-half the historic growth rate of 7 percent per year prior to 1973.

Today almost one-third of our total electricity comes from petroleum (oil and gas). We plan to cut down this contribution to one-fifth as early as 1983; by the year 2000 it should vanish. The reason is obvious. Oil and gas are in short supply and will remain so. They will have to be replaced

---

[8]I hope the reader will not ask why electric generators of 30 percent efficiency should not be replaced right away with those of 50 percent efficiency. One reason is that at present the most efficient plants are only a little above 40 percent. The other reason is that capital investment is particularly high for electricity. The oil crisis has focused our attention on raw materials, and we are apt to forget that fuel is only part of the real expenditure; the other part, capital investment, implies human labor and also more energy. Indeed, a move to replace present equipment with more efficient new equipment without delay would intensify the energy crisis.

by new sources, including cogeneration (not explicitly mentioned in the table), where oil and gas may still make an indirect contribution.

Even today coal plays an important role. More coal can be burned to produce electricity, and with developing pollution abatement more coal can be accepted. This holds for the immediate future because coal plants can be more rapidly deployed than nuclear plants. A further increase in coal plants, however, is undesirable because some pollution from the burning of coal will remain unavoidable.

The main portion of expanded electric generating capacity will have to come from nuclear sources. For 1983 we are considering only nuclear reactors that are already approved, some of which are being erected now. The 4 quads of energy shown in the table could be produced in fewer than a hundred nuclear reactors. The 20 quads indicated for the year 2000 would require approximately 250 reactors. This number compares with the approximately 70 existing reactors, which, however, are on the average somewhat smaller than reactors planned for the future, and it is expected that future reactors will operate at improved efficiency. If we wanted to implement the international option fully, it might be reasonable to erect a great number of reactors and produce more electricity. This energy source could replace more fossil fuels in domestic use, making them available for export. Whether or not this course is reasonable will depend on the price of oil in the international market and the need for petroleum and coal throughout the world.

Hydroelectric power is fully developed in most of the United States, but there are some regions where considerable change may still occur. The most important example is Alaska, where far more electric power could be made available than could presently be used. It may not be too fanciful to imagine that Alaska could develop rapidly and utilize considerably more electric power. There are also interesting minor examples. Perhaps the most beautiful of the Hawaiian Islands is Kauai. Some claim that it has the most rainfall in the United States, or even in the world. Its population will certainly increase, and plans are seriously considered for utilizing its water power. By the year 2000 it may be reasonable to expect that the whole United States will have hydroelectric power production of 5 quads per year.

If we add electricity derived from new technologies, the total figure of 45 quads for electricity by the year 2000 appears to be quite feasible.

Advantages of electricity are its flexibility and cleanliness, though some argue that pollution is only shifted from individual dwellings, shops, or factories, to the central power stations. But pollution is much more easily controlled at these central stations, where emission standards can be enforced, pollutants can be removed, and what remains can be discharged through high stacks so that residual pollution does not affect densely populated areas.

The table apparently shows a decrease in energy devoted to the residential/commercial sector, from 13.6 quads in 1976 to 10 quads in 1983, and to 10 quads in 2000. This decrease is due to the necessary trend toward decreased use of fossil fuels. The total decrease, however, is apparent rather than real. For reasons of bookkeeping we have not included the contribution from electricity in the energy totals. It is unreasonable to do more than simply add the electric energy figures to those directly expended. Instead of the electric energy figures, we should use the equivalent fuel used in generating the electricity. If we do that, the additional energy amounts to another 13.2 quads in 1976, and to additional amounts of 16.2 quads and 19.6 quads in the years 1983 and 2000. This brings the total energy for the residential/commercial sector to 26.8 quads for 1976, 26.2 quads for 1983, and 29.6 quads for 2000. These figures correspond to an initial decrease of consumption from vigorous conservation, and thereafter a consumption increase roughly proportional to the population increase. Considering present trends, this pattern will probably mean a decreased consumption per capita in wealthier portions of our population, not corresponding to a decrease in standard of living but to more efficient use of energy. At the same time it will be possible to increase per capita energy consumption among the poor, a necessary part of raising their standard of living.

In accordance with ideas of the growth option and the international option, the table shows a vigorous increase in the total energy consumed by industries. Contributions from electricity give rise to further acceleration. Apportioning the contributions from electricity and using the same procedures as in the previous paragraph, we find 24.1 quads for industry in 1976, 26.8 quads for 1983, and 44.6 quads for the year 2000. This amounts to a slow but significant expansion of our industrial activities in the difficult years ahead, but to a vigorous 3.5 percent increase per year in the last seventeen years of the century. If substantial energy-saving methods continue, industry will indeed be revitalized.

There is no indication that technology has reached its limits, and there are many ways in which more production can lead to a better life for ourselves as well as for our neighbors. Implementing this industrial expansion will require a great increase in the use of coal and a truly vigorous increase in the use of electricity.

Decreased energy demand is anticipated for transportation, due mostly to more efficient use of gasoline in cars. Such a change is unavoidable, particularly for the next few years, if for no other reason than foreign competition. Taking into account the contribution of electricity, for the year 2000 the total will be 21.9 quads. The increase of energy spent on transportation between 1976 and 2000 would probably keep the per capita expenditure at a constant level. Again, as in the case of use in our homes, energy economy will play a great role, and the actual ability of our people to move around would be increased.

It is hoped that oil will be supplemented not only by electricity but also by synthetic products, 2 quads from biomass, and 1.5 quads from oil shale or possibly coal liquefaction. The last figure is not explicitly shown in the table but is included in the oil supply, which without this help would decrease more sharply in the last years of the century.

Nonenergy applications are assumed to rise sharply in the latter portions of the century because many of these synthetic materials could reasonably replace metals as well as animal and vegetable products. This development would have great potentiality in a rich gas-producing area, which will be mentioned in connection with a worldwide energy model.

In producing synthetic materials, which we call nonenergy, a shift is proposed from oil and gas to coal. Much has been said about coal gasification and coal liquefaction. There is one country where this enterprise is ingeniously pursued: South Africa. Economic success of this South African enterprise is due chiefly to the circumstances that the products derived from coal include valuable synthetics. Once the transformation of coal into other substances is initiated, there are good ways to carry the process to the end of diversified synthetic products, based on the most abundant fossil resource. Correspondingly, we assume that by the year 2000, all synthetic materials in the United States will be derived from coal and biomass rather than from petroleum.

It will be noted that the model will give a rather limited increase in total energy. Actual per capita increase would be approximately one percent a year. Accomplishing this increase without hindrance to industrial growth and to our strength in world relations will require imaginative conservation.

The model overall is based on an optimistic outlook. We assume not only that real progress will be made in new technologies and that nuclear reactors will be rapidly deployed, but also that price ceilings for oil and gas will be removed. What is of immediate importance to energy supply is that, with revival of oil exploration and development, the decline of the gas supply can be limited and the oil supply increased for a few decades. Figures given under Supply are almost as high as the optimistic estimates in the *National Energy Outlook* published by the Federal Energy Administration in 1976. Optimistic figures are adopted because the assumptions on recovery of hydrocarbons left underground have been quite conservative. Tertiary recovery methods for oil and, more generally, *in situ* transformation of fossil fuels (coal or oil shale) can, by the year 2000, make a real contribution to our domestic hydrocarbon supplies.

This point concerning supply is demonstrably crucial. If we can maintain gas production at a somewhat reduced level and increase oil production as indicated, our oil imports will rapidly dwindle and a serious crisis in the natural gas supply can be avoided. Otherwise, there is no escape from continued dependence on foreign resources and from crippling effects of energy shortages.

In the timetable given in the introduction to this volume, five years was the period in which we have to deal with the energy shortage. The model for 1983 was constructed with this time limit in mind. What must be done if, in reality, a solution or a good approximation of a solution is to be obtained by 1983? The most immediate possibility and the most urgent need is energy conservation. In part this means fostering a "conservation ethic." Individually we should use less energy whenever possible. It will take a little longer to conserve by a second method, the use of energy-conserving technology both in our homes and in industry. Conservation was assumed in construction of the model. Without conservation the total energy needed in 1983 for a healthy economy would have been close to 90 quads, corresponding to an annual growth of approximately 3.5 percent in energy consumption, rather than the 76 quads given in the table.

Conservation is necessary but in itself insufficient in the critical area of oil and gas, which together accounted for more than two-thirds of our energy consumption in 1976. One other simple measure is certainly required. We have to deregulate the prices of oil and gas. Both have been kept at artifically low levels. Deregulation would have two effects. One is immediate and small, the other somewhat delayed, but much bigger. The immediate effect is that a price increase will slightly reduce consumption of these essential commodities. Deregulation is actually one of the conservation measures enforced by Adam Smith's ingenious idea that higher prices will decrease demand without the application of any red tape. It will take a little longer for the high price to stimulate production. The present shortage is a consequence of the oil and gas wells we failed to drill a few years ago. Stimulation of more exploration, which will follow a price increase, will increase production slightly in two years, but substantially in five years.

Conservation means to eliminate unnecessary luxuries in energy consumption. We should similarly eliminate unnecessary luxuries in the popular antipollution campaign. In particular, the use of coal has been badly restricted by rigid application of emission standards (which apply equally today whether or not the wind will carry emitted gases away from populous areas) and by overregulating strip-mining without sufficient regard to restoration possibilities. If the coal supply is to increase in a manner comparable with the assumptions in our table, reasonable and reasonably permanent regulation of the coal industry is necessary. Legislation enacted in 1977 will, it is hoped, serve this purpose.

The people of seven states have rejected antinuclear initiatives—in California, Oregon, Washington, Colorado, Arizona, Montana, and Ohio. Despite these votes, obstruction by antinuclear forces continues to slow down deployment of reactors that could generate more electricity. For the near future, the expected increase in nuclear energy generation will supplement the amounts obtained from coal if we remove unnecessary roadblocks. Together, coal and fission account for almost the entire increase in energy production anticipated in the next five years. Without these sources our economy will not only stagnate, but will deteriorate.

It would make a much better story if we could point to one special way for dealing with the energy problem. A single proposal would at least please *some* of the people. To do everything that appears necessary will probably displease everybody. Such a broad approach is, nevertheless, almost certainly the right one.

Let us now turn to the more distant future, and to the wider problem of energy around the world.

By the year 2000 we assume that energy demand in the United States will have reached 106 quads per year. This figure corresponds to a slightly more rapid growth in energy demand than that of the last five critical years. Even so, it amounts to restraints so that the per capita energy consumption would grow only by approximately 1 percent per year. We assume that between 1983 and the year 2000, consumption will increase by

30 quads. Of this increase, 11 quads will come from coal and 14 from nuclear reactors. The two together will account for 83 percent of the growth. Contributions from other sources are important, because oil and gas will probably decline in this period; we assume that this decline will amount to 14 quads. The considerable additional amount of 19 quads comes mostly from novel sources, including 13 quads from various forms of solar energy and 4 quads from geothermal energy. (Two quads come from an increase in hydropower.)

In the next five years we would do well to meet our immediate needs. It is important, however, that in doing so we can eliminate half of our oil imports as early as 1983. According to our model, we should rapidly approach self-sufficiency. In 1976 we imported approximately seven million barrels of oil per day. Reduction of this import is significant to others as well as to ourselves. Seven million barrels per day is one-fifth of the OPEC production. Such a drop in demand for OPEC oil would forestall future price increases. In the long run it might even result in a price reduction.[9]

If we cannot decrease our oil imports soon enough, at least we should switch the source of supply to the Western Hemisphere; we would thus avoid dependence on imports brought by a long and vulnerable ocean voyage. This switch can be done easily because of big oil discoveries in Mexico that make the use of a pipeline feasible. Mexico's projected population increase, with 100 million expected by the year 2000, adds to the desirability of help for that country's economic development. Petrodollars spent south of the Rio Grande may encourage more effective cooperation in North America.

Too much influence in too few hands is rightly unpopular. The power in the hands of Arab oligarchies is a relevant example. The circumstance that this power is of very recent origin and is wielded by unstable and endangered governments does not help; when power is coupled with experience and reliability, it is more easily accepted.

What are the immediate economic and political consequences of the energy shortage? An obvious answer would seem to be that wealth and

---

[9]Unfortunately, events so far are not supporting our model. By the fall of 1978, instead of decreasing, our oil imports increased to eight million barrels a day; furthermore, OPEC has announced a 14.5 percent increase in its oil price. The two events are not unconnected.

power have passed from western democracies to oil-rich underdeveloped nations. This is one part of the picture, of course, and the change has been hailed by the Third World. At last, they thought, the stranglehold of the old colonial nations has been broken.

Actual consequences may be quite different. The oil shortage is disturbing to the advanced democracies, but it affects most of the developing nations in a much more painful manner. They need energy to develop industry. They especially need oil to drive the moderate-sized machines that will generally be widely dispersed. Without oil they will not be developing—they will stay underdeveloped.

But the situation is even worse. In India, water was pumped for a thousand years by "Persian" wheels powered by oxen. The oxen were replaced in the 1960s by oil-powered pumps. Now that oil has become too expensive for India there is trouble in agriculture, and that means the daily bread.

And irrigation is not the only purpose for which energy is needed. Farm machinery in general plays an important role, but energy is most particularly needed for one essential kind of fertilizer without which the green revolution cannot succeed.

Nitrogen-containing compounds are essential to plants. The most abundant source of these compounds today is nitrogen in the atmosphere incorporated into chemical fertilizers by the use of energy derived from oil, and even more preferably from natural gas. The origin of these fertilizers goes back to the days before World War I, when Fritz Haber, in Germany, worked out an industrial method to transform atmospheric nitrogen into ammonia. This process was greatly improved in the 1960s by the Kellogg Company in the United States. Abundant artificial fertilizers have played an important role in our ability to feed the ever-expanding world population. Today this approach runs into the great difficulty of rising energy prices.

The three problems are inseparable: population explosion, food shortage, and energy shortage.

The world today is in the midst of an historic transition. The Industrial Revolution that started in England and spread through Europe, the United States, Japan, and Russia, is now spreading across the whole world; early in the next century it will have engulfed the whole globe. The first result of this scientific-technological revolution was improved health care. DDT has eliminated malaria; better medical services, even in the

most primitive parts of the world, have drastically lowered infant mortality rates. The population explosion deplored by so many is actually a consequence of these beneficial accomplishments.

In view of this fact the population explosion must be considered a strictly temporary phenomenon. Its cause is simply that people have not yet adjusted to the fact that most of the children born now survive. Nor have they adjusted yet to the circumstance that easy birth control is available.

But even though the population explosion is temporary, at this moment it is serious. Today the population of the earth amounts to four billion, and these people are remarkably young. It has been predicted that by the year 2000 there will be seven billion people alive—unless they die of hunger.

One may take hope from some recent predictions that the world population in the year 2000 will be five and a half billion. If this is a valid application of the art of demography, we still shall face difficulties, but not truly desperate ones.

In the near future the problem of food is probably the greatest problem that faces mankind. This is not the place to discuss that problem except insofar as food production is related to energy. The most important consequence of the oil shortage is not that power has shifted from the West back to the Middle East, which is the cradle of western civilization. The most important economic and political consequence of the oil shortage is that it aggravates the food shortage with which we have to deal in the remaining decades of this century.

With this background, let us return to the energy situation in the world. Today world energy consumption is approximately 300 quads. Of these, nearly 200 come from oil and gas, nearly 100 come from coal, and less than 10 from hydroelectric sources; all other contributions are minor. In the year 2000, the world will need 520 quads per year, assuming a population of seven billion, and provided that per capita consumption remains the same. But if there is no more than this 520 quads, then 75 percent of the people will remain miserably poor, just as they are today. Continuation of the great differences in standards of living will probably lead to increasing trouble. Poor people cannot conserve energy. They have much less than what we would consider minimum requirements. While conservation by those who are better off is badly needed, it will provide no more than a few drops in the bucket.

Producing even 520 quads per year will not be easy. It requires op-

timism to predict oil and gas production amounting to 250 quads. This could be accomplished more easily if the massive amounts of gas that are now flared in the developing countries were utilized, preferably to produce fertilizers and methanol. Coal might increase to 200 quads. Hydroelectric power, solar energy, and geothermal energy might give another 70 quads, which just adds up to 520.

There is one ace in the hole: nuclear power. For the United States we have estimated 20 quads from that source. For the world as a whole it may be as much as 100 quads. The increase from 520 to 620 quads, an increase of almost 20 percent of the available energy, might be enough to give people patience to wait for another twenty-five years while further improvements are accomplished.

It is remarkable that in the underdeveloped countries, between 1950 and 1974, per capita energy consumption tripled, while in the developed countries the per capita consumption increased only 1.7 times.[10] In spite of this fact it was said that "the poor get poorer."

There are two favorable factors. The Third World should get a disproportionate share of an energy increase, as has been the case in the past three decades. This could boost their per capita energy consumption by 70 percent, provided the developed world limits its per capita demand for energy.

The second favorable factor is that if the population in the year 2000 is only 5.5 billion, with the reduction of fertility occurring in the developing world, per capita energy consumption in Third World countries could increase by 130 percent. Even with these extreme assumptions, their per capita consumption would be little more than 40 percent of that in the developed world, and less than a quarter of present per capita energy consumption in the United States.

When we speak of nuclear power we must be a little careful. In a strictly technical sense, production equivalent to 100 quads could be accomplished on a worldwide scale, but the main purpose of these nuclear reactors is to produce electricity and to produce it in massive units. Such production is very helpful in advanced countries with a good distribution system for electricity. But for the developing countries, energy is needed

---

[10] In this crude estimate China is included among underdeveloped countries, and Russia among developed countries. See *World Energy Supplies, 1950–1974* (New York: United Nations, 1976).

FIGURE 14-1 *Did you notice the owl (a distant relative) in Figure 4-2? (George Bing.)*

in smaller units and in a more dispersed fashion. It would be reasonable, therefore, to supply the highest possible fraction of energy from nuclear reactors in developed countries. This would free oil for use in the developing countries, in small machines, factories, and other establishments. In no way does this mean that nuclear energy should not be made available in the Third World. It means that the best sequence of utilization would assign greater amounts of fossil fuels for use in early stages of development. In the process of extending the Industrial Revolution over the whole surface of our globe it is reasonable, and to be expected, that fossil fuels will lead the way, and other energy sources will follow.

This model describes a generally optimistic picture that might be drawn for the rest of the century. Can it be done?

The chapter so far contains a few thousand words. Let us conclude it by directing the reader's attention to two illustrations, each of which should be better than a thousand words. The first one speaks for itself. The second depicts the heraldic animal of the establishment which, after all, will have to be entrusted with the execution. Its motto is valid, though a little discouraging: "Let not your right front foot know what your left back foot is doing."

FIGURE 14-2 *Einstein's famous equation* $E = mc^2$ *is, of course, correct. The equation* $E \rightleftharpoons mc^2$ *is a chemical notation meaning that mass and energy can be converted into each other. The formula on the blackboard says that this conversion, in general, cannot be done. If the conversion were possible, a ton of any material would supply the energy needs of the United States for a year. (The centipede was created by Sandi Guntrum).*

EPILOGUE

In the winter of 1977 something peculiar happened in the western hemisphere. This happening may have a decisive effect on whether or not we shall face up to the energy crisis.

Each winter there is a mass of cold heavy air sitting on the pole. The westerly winds in the temperate zone blow over this heavy pool. The line of separation, which we can imagine as a big circle around the pole, is called the cold front or polar front.

When wind blows over water it raises waves; the stronger the wind the longer the waves. Something similar happens when the westerlies blow past the polar front, only there the waves are longer still. On a path around the globe there are approximately four complete waves along the polar front, so that our circle becomes a wavy line.

Eventually the crests grow and break off, and polar air travels south. The polar airmass mixing with warmer moist air brings rain and snow, which are the features of winter weather in the northern temperate zone. This happens every year, and it happened in the winter of 1976–1977. But in that particular winter the Arctic airmass started to move south in the western hemisphere at a somewhat unusual place.

In most winters the cold air starts to move south over Alaska and the Aleutians. Indeed the Aleutians have fabulously horrible winter weather. From there the cold air sweeps over the Pacific, becomes warmer, takes up some moisture, turns west, and dumps rain on the western coast of the United States. In the winter of 1976–1977 the cold air moved south a couple of thousand miles farther east. It started southward over Canada, swept over Chicago, dumped more than ten feet of snow on Buffalo, reached Florida, froze the vegetables and the oranges, and produced the first snow in Miami in recorded history. At the same time, west of the Rockies and over the Pacific, warm air moved north. There was drought in California and a heat wave in Anchorage.

These events brought to a head the energy crisis in the United States. The shivering East and Midwest depleted their scanty gas supply. In California there was far too little rain, further endangering crops and also calling into question hydroelectric power supplies for the summer of 1977.

President Carter got legislation passed permitting him to direct natural gas from intrastate markets in the South (where the relatively high price corresponded to $9 per barrel in oil equivalent) to interstate pipelines, bringing badly needed fuel to Pennsylvania, Ohio, and other hard-hit states. This gas had to be sold at higher prices than had hitherto been permitted in interstate commerce. Thus, responding to necessity, Carter took the first step toward deregulating gas.

In order to keep people from freezing, industries had to be shut down, and unemployment rose temporarily by more than a million, demonstrating the impact of energy shortage on the economy. The president wore a sweater in his first fireside chat, the fire being particularly welcome at that time. He had planned to emphasize conservation, and he did so with understandable fervor.

The hard winter of 1976–1977 will be a memory when this book reaches the reader. Will the memory be well remembered or will it be forgotten, as the oil embargo was forgotten as soon as the lines of cars at the gas pumps disappeared? At a moment of need the president took a step toward deregulating gas. Will that step be followed by permanent deregulation? Will we then take all the necessary and often painful steps that are needed to deal with an energy crisis that might better be called a chronic energy disease?

The United States is in a pivotal position. It has the natural resources and the industrial capability to do something in the healing process needed to recover from the energy disease. There is hope that the shock of the winter of 1976–1977 will start a healthy process of recovery. Realization is growing in the United States that the nation's fate and the fate of the rest of the world are closely related. From this realization it is only a short step to the understanding that the energy problem must be dealt with, not exclusively from the point of view of the United States, but with a view to the full contribution that this country can make to the world's energy supply.

In Chapter 14, I described an energy model that I do not expect will actually correspond to what is going to happen. The figures in the model are not to be taken as factual data; I hope they will serve as bases for debate. The model does not demonstrate what really will happen by the year 2000. But by that date it should be possible for the United States to make substantial contributions to the world economy in the field of energy.

Today more than two-thirds of all the grain in the international marketplace comes from the United States. If the United States could supply energy as well as food, and this possibility by the year 2000 is by no means excluded, then our contribution to world stability would be so great that no one could ignore it.

Since I have moved in the preceding paragraphs and in this epilogue to the brink of making predictions, I shall conclude with a prophecy which is explicit but ambiguous. In the year 2000 the state of the world will be either very much better or very much worse than it is today. The doubtful balance we see today will not last for many more years. It depends on decisions made in the near future whether life on our planet will be regulated and grim or whether it will be consistent with human dignity.

INDEX